EXTREMELY HARDCORE

INSIDE ELON MUSK'S TWITTER

Zoë Schiffer

PORTFOLIO / PENGUIN

Portfolio / Penguin
An imprint of Penguin Random House LLC
penguinrandomhouse.com

Library of Congress Cataloging-in-Publication Data

Names: Schiffer, Zoë, author.
Title: Extremely hardcore : inside Elon Musk's Twitter / Zoë Schiffer.
Description: New York : Portfolio/Penguin, [2024] |
Includes bibliographical references.
Identifiers: LCCN 2023048761 (print) | LCCN 2023048762 (ebook) |
ISBN 9780593716601 (hardcover) | ISBN 9780593716618 (ebook)
Subjects: LCSH: Musk, Elon. | Twitter (Firm) |
Online social networks—United States. | Internet industry—United States.
Classification: LCC HM743.T95 S35 2024 (print) | LCC HM743.T95 (ebook) |
DDC 302.30285—dc23/eng/20231019
LC record available at https://lccn.loc.gov/2023048761
LC ebook record available at https://lccn.loc.gov/2023048762

Printed in the United States of America
1 3 5 7 9 10 8 6 4 2

BOOK DESIGN BY NICOLE LAROCHE

Some names and identifying characteristics of people have been changed to protect their privacy.

EXTREMELY HARDCORE

For my dad.
Still the best writer I know.

Three sparks that set on fire every heart
are envy, pride, and avariciousness.

DANTE'S *INFERNO*, CANTO VI

IF THE ZOO BANS ME FOR HOLLERING
AT THE ANIMALS I WILL FACE GOD AND
WALK BACKWARDS INTO HELL

@DRIL

CONTENTS

Introduction: "This App Makes Zero Fucking Sense" xiii

Part I

THE BIRD APP

1. "This Is Actually Me" 3
2. "Benevolent Anarchy" 10
3. "Little Tentacles" 19
4. "Be There, Will Be Wild!" 23
5. "Woke Mind Virus" 25
6. "Welcome Elon!" 30
7. "Midas Touch" 38
8. "Back Door Man" 41
9. "I Made an Offer" 45
10. "I'd Jump on a Grande for You" 49
11. "Let's See What This Guy Can Do" 54
12. "Temporarily on Hold" 56
13. "Why We're Here" 58
14. "Hide-and-Seek" 61
15. "Some Things Are Priceless" 65

Part II

HELLSITE

16. "Let That Sink In!" 75

17. "I Understand How Computers Work" 82

18. "Content Moderation Is a Product" 86

19. "The Bird Is Freed" 89

20. "We Truly Cobbled It Together" 94

21. "Comedy Is Now Legal on Twitter" 96

22. "Please Print" 98

23. "I Bet My Emergency Is Bigger Than Yours!" 100

24. "Should I Post It?" 105

25. "#SleepWhereYouWork" 108

26. "'Maniacal' Urgency" 112

27. "Uncertain Times" 114

28. "Your Role at Twitter" 116

29. "Weak, Lazy, Unmotivated" 122

30. "New Blue . . . Coming Soon!" 125

31. "The Economic Picture Ahead" 133

32. "Elon Puts Rockets into Space, He's Not Afraid of the FTC" 135

33. "Resignation Accepted" 137

34. "High Risk" 142

35. "How It Started/How It's Going" 145

36. "Extremely Hardcore" 149

37. "Vox Populi, Vox Dei" 156

38. "Our Concerns Regarding the Quality of Your Coding Ability" 160

39. "Exceptionally Poor Taste" 162

40. "This Will Be Awesome" 166

41. "Assassination Coordinates" 177

42. "Epic Shit" 182
43. "This Explains a Lot" 184
44. "Deep Cuts" 188
45. "We Just Won't Pay Those" 191
46. "An Absolute Scam" 196
47. "#TwitterDown" 199

Part III

MAIN CHARACTER

48. "One Main Character" 207
49. "VIP Users" 210
50. "What the Fuck Is Going on with the App" 212
51. "Engagement Night" 214
52. "These People Are Pillaging Us" 220
53. "Caught Red-Handed" 223
54. "I'm Not Going to Bullshit You" 227
55. "Official Company Communication" 230
56. "A Recently Fired Twitter Employee" 232
57. "Good Thing I Fired Him" 233
58. "Inverse Startup" 237
59. "You're Welcome Namaste" 240
60. "Try It, but Don't Trust It" 243
61. "Did Your Brain Fall Out of Your Head?" 246
62. "I Need to Do Some Deleting" 249
63. "The Same Rules & Rewards" 253
64. "Every Parent Should Watch This" 256
65. "Someone Foolish Enough to Take the Job" 260
66. "Serious Numbers" 265

67. "Sorry, You Are Rate Limited" 268
68. "I'm Up for a Cage Match" 273
69. "We Shall Bid Adieu to the Twitter Brand, and Gradually, All the Birds" 278
70. "I Wish I Had Been More Worried" 281
71. "The Stakes" 283
72. "Zero Sum" 285

Conclusion 293

ACKNOWLEDGMENTS 297
NOTES 299

INTRODUCTION

"This App Makes Zero Fucking Sense"

On February 12, 2023, Elon Musk sat on his private jet, fuming. He was flying home from the Super Bowl in Glendale, Arizona, but his mind wasn't on the game. Earlier that day, both he and President Joe Biden had tweeted their support for the Philadelphia Eagles. But according to Twitter's engagement metrics, Biden's tweet had three times the number of views. *What the hell?*

Four months earlier, Musk had acquired Twitter, making him not just the social media platform's most powerful figure, but also its most ubiquitous. He posted constantly—recycled memes, missives about free speech, promises about upcoming features. Day in and day out, he was the indisputable main character of Twitter.

Then, in early 2023, Musk's engagement started tanking. The richest man on Earth simply couldn't fathom why. His photos of rockets were awesome. His jokes were never *not* funny. Plus, he had more followers than anyone else—and nearly a hundred million more than @POTUS. How could he lose to a damp sock puppet in human form who happened to be president of the United States? Musk had already called multiple meetings to demand answers from Twitter employees. "Jesus H. Christ," they'd heard him muttering. "This app makes zero fucking sense."

As the Eagles and Kansas City Chiefs traded touchdowns at State Farm Stadium, the engineering team at Twitter's headquarters in San Francisco scrambled to come up with an answer. Musk suspected foul play. Had a spiteful employee planted a bug in the algorithm to suppress the Like count on his posts?

One week earlier, one of Twitter's highest-ranking engineers had dared to say what many understood to be obvious: that the drop in engagement was organic. "If you look at Google Trends, interest in your name is on the decline," the engineer, Yang, told Musk. He showed Musk a graph with an impressive spike in April—when he'd first announced his plans to buy the platform. It was followed by a jagged downward slope. Interest had gone from a score of one hundred to a score of just eight.

"You're fired, you're fired," Musk hissed. Yang walked out. Then Musk turned to the rest of the team. "This is ridiculous. I have more than a hundred million followers, and I'm only getting tens of thousands of impressions," he said, according to three employees who were present. No one said a word.

"Why is nobody else here speaking?" Musk said, sounding exasperated. He told the group they'd reconvene the next day. If he didn't get a straight answer, they'd all be fired.

After that, no one else tried to challenge Musk's reality.

Musk's fraught takeover of Twitter had captivated the country for months. The genius behind Tesla, SpaceX, The Boring Company, and Neuralink had grandly declared that his next mission was to restore free speech to the public town square. "This is a battle for the future of civilization," Musk tweeted in November 2022. "If free speech is lost even in America, tyranny is all that lies ahead." But now, in early 2023, after months of firing staffers, banning journalists, and twisting content policies into pretzels, the acquisition increasingly looked like a vanity project. Twitter had never been profitable on the scale of its competitors (it made a modest $5.08 billion in 2021), but now its revenues were collapsing, down 40 percent from the year before. Advertisers had fled the platform. The circle of people who saw the billionaire as a visionary was shrinking—not that the billionaire seemed to realize it.

After Musk's jet touched down in Oakland, his cousin, James Musk, jumped into action. The Tesla Autopilot engineer had joined Twitter the previous October to help usher the company into its new era. "@here we are debugging an issue with engagement across the platform," James wrote cryptically on Slack at 2:36 a.m. "Any people who can make dashboards and write software please can you help solve this problem. This is super high urgency. If you are willing to help out please thumbs up this post."

One of Twitter's core values had been "defend and respect the user's voice." Now, the only voice that mattered was Elon Musk's.

Many employees viewed the late-night demand as a desperate attempt to placate an insatiable ruler. But Randall Lin, a machine-learning engineer, knew it was also an opportunity. Lin's job was to make Twitter's home timeline as relevant and engaging as possible. He didn't have to drop everything to prioritize Musk's urgent project—he *wanted* to. Musk was mercurial; a high achiever like Lin could easily rise up the ranks if he played his cards right. "Everything else you are focusing on is great," James had told him before the Super Bowl, "but there is nothing else on Elon's mind but the engagement issue."

Lin and around eighty colleagues worked through the night rewriting the Twitter algorithm. First, they applied a special signature to Musk's profile to ensure he showed up in almost every user's feed, whether they followed him or not. Then they applied a "power user multiplier" to artificially boost his tweets by a factor of one thousand.

The next day, a Monday morning, Twitter users logged on to see an entire feed of Elon Musk. His replies to obscure right-wing accounts were showing up at the top of the app. People were furious. "why the absolute fuck is elon musk all over my for you on twitter?" asked Twitter user @kenminkim. "My 'For You' page is literally just Elon Musk replies and ads lmao," wrote @TayInLA_. "Is Twitter literally just his personal mouthpiece now?" asked @johnjsills.

The outrage made Musk more ecstatic. He roamed the halls of Twitter HQ, thumbing through his feed, delighted. "It's just like that meme of that girl pouring milk down her friend's throat," he told employees happily, shaking his head. "That's like me with the tweets." Moments later, he tweeted the meme, labeling the blond girl pouring milk as "Elon's tweets" and the brunette being force-fed "Twitter."

Just fourteen months earlier, *Time* named Elon Musk the Person of the Year. "This is the man who aspires to save our planet and get us a new one to inhabit: clown, genius, edgelord, visionary, industrialist, showman, cad; a madcap hybrid of Thomas Edison, P. T. Barnum, Andrew Carnegie and *Watchmen*'s Doctor Manhattan, the brooding, blue-skinned man-god who invents electric cars and moves to Mars," the magazine wrote.

Musk had recently asked his 62.8 million Twitter followers whether he should sell 10 percent of his Tesla stock (and raise his taxable income). The richest man in the world had grown tired of getting criticized for not paying enough taxes. His followers thought it was a good idea: 57.9 percent voted yes.

The sale left Musk with $10 billion in ready cash. He could have bought fifteen private islands. He didn't. He started buying up Twitter shares.

Musk offered to buy the company and take it private in April 2022, then changed his mind as the stock market tanked, then changed it back. By the time the deal closed in October 2022 for $44 billion, including a $24 billion investment from Musk himself, he'd overpaid by roughly $19 billion, according to one credible analysis.

Many commentators were excited to see what Musk would do. "Elon Musk is an amazing entrepreneur, an extraordinary innovator. He's the

Henry Ford of our time," said Jonathan Greenblatt, CEO of the Anti-Defamation League. "He's taken on big huge complex tasks that no one thought could be solved like rocketry or mobility or solar . . . And to think what he can do with the public square."

Musk galloped into Twitter's sleepy headquarters like a conquering general, wasting no time in cutting costs and employees. "Going forward, to build a breakthrough Twitter 2.0 and succeed in an increasingly competitive world, we will need to be extremely hardcore," he wrote in an internal email on November 16, 2022. "This will mean working long hours at high intensity. Only exceptional performance will constitute a passing grade."

If there was a unifying philosophy at play, it was rage against the "woke mind virus." Anything that was "anti-meritocratic and anything that results in the suppression of free speech" was a menace. In April 2022, when Netflix shares cratered, Musk blamed wokeness for making the streaming service unwatchable. The stakes were even higher at Twitter. To Musk, the future of democracy was on the line. Leftist content moderation, office lattes, twenty-week parental leave, conservative "shadowbanning," holiday breaks, regular janitorial services: all artifacts of Twitter's woke culture that needed to be uprooted.

Other tech leaders were watching.

"I guess the times of complaining to the CEO of a large tech company at an all hands in front of thousands of people about the quality of toilet paper have come to an end. (True story. This really happened.)," former Meta executive David Marcus tweeted shortly after Musk bought Twitter.

Three months after the deal closed, Mark Zuckerberg announced Meta was in its "year of efficiency," after shutting down experimental projects and laying off 13 percent of its workforce. He later credited Musk's example with giving him license to ruthlessly fire middle managers.

At first, it seemed obvious to many that Musk would succeed. "Watching @elonmusk + Co take over Twitter is like watching a business school case study on how to make money on the internet," tweeted *The Information* founder Jessica Lessin.

Except that's not how it played out. The attributes that made Musk good at tweeting—a combination of recklessness and shamelessness—made him exceedingly bad at running Twitter. His impulsiveness did not play well with advertisers. His thirst for speed alarmed regulators. And his opposition to content moderation alienated regular users. Six months after the deal closed, Twitter had lost two thirds of its value and found itself in hot water with lawmakers in the United States and Europe.

As Twitter's business went into free fall, Musk's reputation took a hit. To many, he'd become more edgelord, less visionary. "Content moderation is really hard and apparently harder than rocket science," noted Evelyn Douek, an assistant professor at Stanford Law School, who studies online speech regulation.

This book charts Musk's journey to acquire Twitter from January 2022, when he began quietly buying shares, to October 2023, one year after the deal closed. By this time, Twitter had rebranded to X, and the bird app was a piece of tech history. You'll meet a machine-learning savant who went all in on Twitter 2.0, a father trapped in his job because his kids needed his health care, a venerated infrastructure engineer who was fired after standing up to her boss *on Twitter*, a trust and safety advocate who became the subject of a life-threatening harassment campaign, and dozens of employees who lost their jobs trying to save the company they helped build.

This is another way of telling you, right up front, that this book isn't a biography of Elon Musk. Nor is it a story about his version of events, which he's been live-tweeting since the saga began. It is, instead, the story of what happens when the world's richest man walks into the lobby of your workplace holding a kitchen sink. It is about the collapse of morale after he lays off the majority of the workforce. It is a cautionary tale about the ripple effects on politics and culture when a billionaire provocateur meddles with a massively influential institution he doesn't care to understand. And it is a chronicle of workers who tried to stop a renegade billionaire from upending online speech in one of the greatest unforced errors in Silicon Valley history.

The following chapters are the result of hundreds of hours of interviews with more than sixty employees at all levels of the company, from the C-suite to the front lines, as well as dozens of outside experts, over the course of fourteen months of reporting, along with hundreds of pages of internal memos, whistleblower complaints, and court documents. Some sources worked alongside Musk for months; others never met him but felt the collateral damage of his leadership. Dialogue has been re-created from recordings and the recollections of multiple sources. The reporting draws from years of work between me and my colleague and mentor Casey Newton, who founded the investigative tech newsletter *Platformer*, and my time as managing editor at *Platformer* and as a senior reporter at *The Verge*. While I contacted Musk multiple times to ask for interviews, he responded only once, with a cry-laughing emoji.

Many sources agreed to use their real names—at great personal risk to their jobs and families. Others remain anonymous out of fear of professional retribution. For a story like this, where the balance of power lies squarely with upper management, I believe this anonymity is warranted.

It's a business case study, a labor investigation, and a murder mystery. The body count is still being tallied. And if there's one thing Musk is right about, it's this: the story is *extremely hardcore*.

Part I

THE BIRD APP

CHAPTER 1

"This Is Actually Me"

Elon Musk's first tweet, sent on June 4, 2010, established that he was the real thing. "Please ignore prior tweets, as that was someone pretending to be me :)," he wrote. "This is actually me." Like many celebrities, he'd been impersonated by parody accounts on Twitter. But thanks to the company's verification system, users would now know that this Elon Musk, with the handle @elonmusk, was the genuine artifact.

Not that there was much to follow in the early days of Musk's time on the platform. Like many people new to Twitter, it took him a while to find his footing. His early posts were mostly banal—things he was reading, recommendations of stuff he liked. Once he got going, he was hooked. Social media gave him a way to talk to his audience, circumventing traditional media outlets. Now, if Musk wanted to hype the accomplishments of his visionary rocket company SpaceX, or his pioneering electric vehicle company Tesla, he wouldn't need to do it through a gatekeeping journalist. He could just pull out his phone and speak directly to his followers.

Musk made his first millions in 1999, when Compaq bought his startup Zip2, a yellow pages–like local directory and guide. (His father, Errol Musk, had provided the company with $28,000 in start-up money.) After the sale, he purchased a million-dollar sports car, the McLaren F1. CNN was at Musk's house to film its arrival. "Some could interpret purchasing this car as behavior characteristic of an imperialist brat," Musk, a white South African, said matter-of-factly. He was wearing an ill-fitting brown sport coat, his hair noticeably thinner than it is today. "Just three years ago, I was showering at the [YMCA] and sleeping on the office floor, and now obviously I've got a million-dollar car and quite a few creature comforts," he added.

Musk poured more than half his earnings into an ambitious enterprise called X.com, an online financial services startup. In 2000, X.com merged with Confinity, a rival payments company cofounded by Peter Thiel. Musk became the CEO of the company that would later be known as PayPal.

That March, Musk took Thiel for a ride in the McLaren. "So what can this do?" Thiel asked, as they drove up Sand Hill Road in the heart of Silicon Valley. "Watch this," Musk replied. He rammed his foot onto the gas, changed lanes, and hit an embankment. The car shot up in the air and started spinning. When the McLaren landed, fortuitously pointing in the right direction, the impact less fortuitously shattered the windows. "This isn't insured," Musk told Thiel. He repaired the car and said he sold it for a profit years later, cementing the story as a beloved part of his personal mythology. Musk was a renegade, but things always worked out for him in the end.

That same year, Musk was pushed out by the board of the newly combined company. One of the major disputes was his insistence on calling the entity X. As he stepped on a flight for his long-postponed Australian honeymoon with his first wife, Justine Wilson, executives delivered a letter of no confidence to the board. The coup had been partially orchestrated by Thiel, who would later become a venture capitalist and a shadowy power broker in a right-wing war against journalists.

Musk had no choice but to exit. In the years to come, as the internet became ubiquitous and mobile, he mused about his vision for X.com. He didn't just want to revolutionize banking—he wanted to create an everything app. He would cling to this idea for two more decades, nursing his obsession with the letter X.

His experience at PayPal humbled him. But it also made him even richer. In 2002 the company was acquired by eBay for $1.5 billion in stock—with Musk, its largest shareholder, netting between $160 million and $180 million. Musk put that money into an even more challenging idea: spaceflight. More specifically, he believed that human beings should colonize Mars. He would start with rockets, picking up the work that NASA was no longer funding with the imagination it had in his youth. Lucrative government contracts were there for the taking. Eventually, Musk planned to make those rockets reusable—a feat never accomplished by any space program.

Though the dream of X never died, it was easy to move on after his ousting. After all, the concerns of PayPal were absolutely terrestrial. Musk was headed to space.

This ambition set Musk apart. Silicon Valley was filthy with founders developing websites, software, and services. But how many people wanted to build rockets?

"He was just another tech boy," veteran tech journalist Kara Swisher tells me. "But then when he started doing the space stuff and the car stuff, that was really interesting, that's substantive. He could've done a lot of other things, and he didn't."

Just ten years after its founding, SpaceX had made great progress.

"Splashdown successful!! Sending fast boat to Dragon lat/long provided by P3 tracking planes #Dragon," Musk tweeted on May 31, 2012. Behind the jargon was a historic announcement: SpaceX had successfully delivered cargo to the International Space Station, making it the first private company to ever do so.

The post garnered only 340 retweets, a number that would pale in comparison to the attention he would later command on Twitter. But at the time, it was enough for Musk. The press was coming to him for comment. Now he was the gatekeeper—the one in control of the story. By the time he acquired Twitter, Musk had tweeted more than nineteen thousand times over thirteen years, roughly four times a day.

The electric car manufacturer Tesla is so strongly associated with Elon Musk that it's easy to forget that he didn't start it. Like SpaceX, it was an ambitious company, attempting to combat climate change—save the planet, and thus, the human race—by shifting the world from its reliance on fossil-fuel-chugging vehicles. It would also do it by making sports cars.

Musk was involved early on as an investor, pumping tens of millions of dollars into the startup. By 2009, of the $187 million Tesla had raised in venture funding, Musk had contributed over a third. At that point, he had already ousted one of Tesla's original founders, Martin Eberhard, who later sued and settled with Musk. But the drama didn't matter. This time, Musk was the ouster, not the ousted.

If people were curious about SpaceX, they were rabid about Tesla. And the best place to stay up-to-date on the company? The CEO's Twitter feed. By 2018, eight years into his tweeting career, Musk knew how to wield the platform. Eventually, he dissolved Tesla's entire PR team in North America. In SEC filings, Tesla's board of directors urged investors to check Musk and Tesla's Twitter feeds for information about the company.

Musk stood out from other CEOs and celebrities on Twitter by personally replying to his followers. Beyoncé's fans could tweet at her all day and she'd never acknowledge their existence. Tweet at one of the richest men in the world and he might actually tweet back. Musk replied to a lot of people, enough to make him relatable, even likable. For a guy worth $19.9 billion, Musk was a man of the people.

The relationship between Musk and his fans was mutually beneficial. In 2020, Tesla was the fifteenth most-held stock on Robinhood, the favored app of retail investors, with more than a hundred fifty thousand individual shareholders.

"If you look at the history of Tesla, I think most people think of Tesla as a pretty good business now," said finance journalist Matt Levine in an interview with Kara Swisher in October 2022. "But for a long time, the stock price was *way* ahead of the business, it would not make money year after year, and the typical reason that people give for that is there's an army of Elon Musk fans who will buy the stock and push up the price of the stock."

Musk eschewed traditional forms of advertising for his car company—an avenue where his biggest competitors, like Ford, Chevy, and Toyota, spent collectively tens of billions of dollars a year. In 2020, Tesla became the world's most valuable automaker. Over the next two years, Musk's net worth shot up from $24.6 billion to $219 billion, vaulting him ahead of Jeff Bezos to become the richest man in the world.

While you could attribute some of the company's success to Musk's Twitter presence, his tweets had gotten him into an awful lot of trouble, too.

In 2018, a group of Thai boys and their soccer coach went exploring in Chiang Rai province and got trapped by a flash flood in a cave deep underground. The mission to save them stretched on for nearly two weeks, captivating the world as the kids' lives hung in the balance.

The saga had nothing to do with Musk or his companies, but when a Twitter user asked if Musk could help, the CEO couldn't help but to accept. "I suspect that the Thai govt has this under control, but I'm happy to help if there is a way to do so," he tweeted, a week after the boys disappeared.

Musk offered to build a mini-submarine to help navigate the caves, and an escape pod to transport the children to safety. He started tweeting out updates. "Construction complete in about 8 hours, then 17 hour flight to Thailand," he wrote on July 7.

He even directed employees to pressure Thai government officials "to make complimentary public statements about him and the technology his engineers were developing to aid in the cave rescue," according to CNBC's Lora Kolodny.

On July 8, 2018, the kids were rescued by professional divers, with no help from Musk.

Vernon Unsworth, one of the experts involved in the rescue mission, went on CNN and called Musk's efforts a PR stunt. "It had absolutely no chance of working," he said. "He can stick his submarine where it hurts."

Musk was furious. "We will make one of the mini-sub/pod going all the way to Cave 5 no problemo. Sorry pedo guy, you really did ask for it," he shot back in a since-deleted tweet.

It was behavior unbecoming of anyone, let alone the CEO of multiple companies. After another Silicon Valley investor asked what he meant by "pedo guy," Musk doubled down, replying, "Bet ya a signed dollar it's true."

Unsworth sued for defamation, but lost in court. (Musk's defense was that "pedo guy" was a generic insult—"I also did not literally mean that he was a pedophile," he said on the stand. "I meant he was a creep.") Unsworth's lawyer on the case was L. Lin Wood, one of the attorneys who tried to overturn the results of the 2020 election and was later permanently banned from Twitter.

As embarrassing as the submarine incident was, many of Musk's fans brushed it off. This didn't have anything to do with Tesla.

Then, in 2018, Musk's tweets got him in trouble again. "Am considering taking Tesla private at $420. Funding secured," he said. He didn't have funding secured. Setting the share price at the number associated with marijuana made the tweet seem like an inside joke. The Securities and Exchange Commission wasn't laughing and sued him for allegedly misleading investors. Musk settled, but it cost him and Tesla $20 million each in fines. He could remain CEO, but he would be ineligible to be elected chairman of the board for three years. Arguably, it was the single most monetarily damaging tweet of all time. (Four months after the infamous funding tweet, in a rare on-camera interview on *60 Minutes*, Musk said, through tears, that he did not respect the SEC.) As a result of the settlement, Musk was also required to have his tweets approved by a "Twitter sitter," a securities lawyer at Tesla, if they contained material business information about the automaker.

If there were lessons to be learned from tweeting indiscriminately, Musk did not care to learn them. In fact, his posts soon became even more inflammatory. In 2020, he denied the seriousness of the Covid-19 pandemic ("The coronavirus panic is dumb"), and shared transphobic beliefs ("Pronouns suck"), cultivating fandoms on right-wing corners of the internet.

The successes of SpaceX and Tesla buoyed his reputation. Sure, he was posting in a manner that would get most other CEOs ousted, but his companies were innovative—and Tesla's share price continued to reflect that. SpaceX was pushing the United States forward in an industry known for groundbreaking science and technology; Tesla had made electric cars a commercial reality. This brilliant technologist, innovator, and businessman could tweet whatever and whenever he wanted—"At least 50% of my tweets were made on a porcelain throne," he claimed—as long as his companies continued to satisfy investors.

If the stock didn't sink, then neither would Musk.

CHAPTER 2

"Benevolent Anarchy"

Jack Dorsey strolled down a white sand beach in Hawaii, wearing a St. Louis Cardinals hat that looked like it had seen better days. It was November 10, 2020, and regular people were stuck at home waiting out the pandemic. But Dorsey was far from a regular person. He was the CEO of Twitter.

Dorsey launched Twitter in 2006 with cofounders Biz Stone, Evan Williams, and Noah Glass. In the early days, the platform was full of promise—the buzziest of the buzzy social apps. While Facebook allowed people to connect, Twitter encouraged them to *follow*, pioneering a one-way social graph that attracted scores of high-profile users.

Twitter had its hooks in people from the start. Dorsey, Stone, and Williams debuted the app at SXSW in Austin, Texas. Rather than give a convention center–sanctioned presentation about what they were building, the group spent $11,000 to display tweets on flat-screen TVs along the indoor walkways. Suddenly, if you wanted to know what people were talking about at the conference, you had to be on Twitter. A *Wired* headline declared: TWITTER IS RULING SXSW.

As the app's influence spread beyond the tech industry, Twitter's user base became a who's who of power players in politics, sports, and media. Barack Obama used it in 2008 to organize a formidable grassroots campaign; Middle Eastern dissidents leveraged it in 2010 to protest authoritarian dictatorships.

The platform played a leading role in high-stakes political dramas, like the raid that killed Osama bin Laden. On May 1, 2011, two dozen Navy SEALs descended on a compound in Abbottabad, Pakistan, and killed the head of al-Qaeda. The top-secret mission was live-tweeted by an unwitting IT consultant who was annoyed by the noise from the helicopters.

"Helicopter hovering above Abbottabad at 1AM (is a rare event)," he said. Eleven minutes later he added: "A huge window shaking bang here in Abbottabad Cantt. I hope its not the start of something nasty:-S."

Political influence became a powerful recruiting tool for Twitter. Employees, who called themselves "tweeps," believed in the company's mission to improve the health of public discourse.

Yet for all its potential, Twitter's business never fully took flight. Cultural influence didn't translate into universal adoption. Twitter had 192 million daily active users in 2020, compared to Facebook's 1.84 *billion*. Its revenue that year was less than 5 percent of Facebook's. From 2006 to 2016, the company failed to turn a profit. Roughly 90 percent of the company's revenue came from ad sales, but it lacked the robust targeting capabilities of Facebook and Google.

The issue was that Twitter rarely shipped new products. Even small design changes required weeks of meetings with numerous stakeholders at seemingly all levels of the company. TikTok and YouTube had better creator tools. Google and Facebook had a better ads stack. Twitter slept on clear revenue opportunities. In 2012, the company purchased Vine, a short-form video app that was an early predecessor to TikTok. It failed to integrate Vine's technology into its tech stack and shuttered the app five years later.

Internally, employees blamed Jack Dorsey. Companies are reflections of their leaders, and Dorsey was an absentee landlord. In addition to running Twitter, he helmed the more successful payments company Block, then known as Square. He lived part-time in Big Sur and talked about moving to Africa, popping up only occasionally to issue gnomic pronouncements about his desire to turn Twitter into some sort of decentralized protocol, to free it from the constraints of Wall Street.

Twitter's ads-dependent business struggled amid the company's ongoing battle with harassment. The very attributes that made the platform addictive—the ability to quote-tweet and see conversations from accounts outside one's personal network—made it ripe for a pile-on. Like most social platforms, Twitter was optimized for engagement, and watching people fight was very engaging, even if it made advertisers wary.

The problem was especially bad for women and people of color. For years, the Black community played a crucial role in Twitter's rise, but the platform did little to tamp down racist content. Until 2016, Twitter had no Black board members. Two years later, a study conducted by Amnesty International found that one out of every ten tweets mentioning Black female politicians or journalists was abusive or problematic.

The situation had financial repercussions. In 2016, Disney pulled out of talks to acquire Twitter, citing the platform's problems with abuse. "You have to look, of course, at all the hate speech and potential to do as much harm as good," CEO Bob Iger said. ". . . This was just something that we were not ready to take on."

The toxicity of the platform was best exemplified by one of Twitter's most notorious power users: the reality TV star Donald J. Trump. In 2015, when Trump announced he was running for president, few journalists took him seriously. How could a washed-up real estate mogul beat former secretary of state Hillary Clinton to become the forty-fifth president of the United States? But Trump understood what the mainstream press did not: attention—not accuracy—was the coin of the realm. Throughout his campaign, Trump dominated the news cycle, gripping the nation in 140-character increments. His popularity swelled—as did his Twitter following. In 2016, when Trump became president, he credited Twitter and Facebook with helping him win.

The 2016 election prompted a reckoning inside Twitter. The Russian government–linked Internet Research Agency had used the platform to spread pro-Trump propaganda, reaching millions of people and potentially impacting the vote. Political polarization was at an all-time high—and Twitter seemed partly to blame. Suddenly, working at Twitter wasn't cool, it was embarrassing.

Employees demanded that the company double down on content moderation to balance newsworthiness against the threat of potential violence. Trump wasn't a regular user, they argued, he deserved—he required—special treatment.

In 2019, Dorsey sat onstage at the TED conference in Vancouver to address Twitter's issues head-on. He was dressed all in black, with a beanie, nose ring, and scraggly beard that made him look a bit like a cult leader. "We have seen abuse, we have seen harassment, we have seen manipulation, automation, human coordination, misinformation," he said ponderously. "So these are all dynamics that we were not, uh, expecting 13 years ago when we were starting the company, but we do now see them at scale . . . It's a pretty terrible situation."

The stammering speech was well-timed. Public opinion on big tech companies was turning. Facebook was still reeling from the Cambridge Analytica scandal, having allowed political consultants to obtain private information on tens of millions of users ahead of the 2016 election. Twitter might not have been as innovative or as lucrative as its competitors, but at least Dorsey was *trying* to fix the problem.

In February 2020, activist investor Elliott Management acquired a 5 percent stake in Twitter with plans to oust Dorsey as chief executive, citing the company's poor management, slow growth, and inefficient product organization. Twitter employees rallied around Dorsey, tweeting their support with the hashtag #WeBackJack. A month later, Twitter struck a deal with the firm. Dorsey remained CEO and Elliott Management got a seat on the board, which created a temporary committee to evaluate

Twitter's leadership structure. The company announced a set of ambitions, including growing its monetizable daily active users by at least 20 percent.

By the time he'd landed in Hawaii, soon meandering down a secluded beach with the actor Sean Penn and Israeli venture capitalist Vivi Nevo, Dorsey was enjoying more support from his workforce than he had in years. Still, his position at Twitter was precarious.

Yoel Roth was on a date when he first realized he wanted to work in trust and safety. It was 2011, and he was getting drinks with a man who worked at the parent company of Manhunt, one of the first gay hookup sites. At the time, the website permitted users to send certain types of nude images, but there were limits to what the rules allowed. "I remember asking, 'How do these decisions get made?'" Yoel tells me. His date responded: "There's a team of people who review every single photo—and most of them are straight women." Roth was floored. "God, there's a team of heterosexual women who have to look at the depraved things that gay men post on the internet?" he recounted in a later interview.

That knowledge led him to consider deeper questions about how social platforms shape the ways that people engage with one another online. A few years later, Roth got his PhD in communication from the University of Pennsylvania, with a focus on social media platform governance.

In 2015, Roth landed his dream job at Twitter. The platform was at the height of its influence and just beginning to grapple in earnest with free speech and content moderation, particularly in parts of the world that lacked free speech protections. Roth sat at a small desk in the San Francisco office, in front of a life-size cardboard cutout of Justin Bieber, who was one of the most popular users on the site.

The job was every bit as fascinating as Roth had imagined it would be. Twitter was fighting troll farms, state-sponsored disinformation, and organized harassment. To most executives, it was a headache; to Roth, it was a satisfying puzzle with no simple solution. He'd spend hours debating what constituted hate speech and when to add a label on misinformation.

Then, in 2020, Roth got his most consequential assignment to date. President Donald Trump, enraged about the possibility of losing his reelection campaign to Democratic nominee Joe Biden, started spreading lies about mail-in ballots, incorrectly claiming that they increased the risk of voter fraud. The move was straight out of Trump's 2016 playbook, where he'd accused his opponent Hillary Clinton of a vague techno-conspiracy involving her emails. At the time, Twitter hadn't done much to stop him. Now, Trump's tweets clearly violated the platform's rules on election misinformation. The most powerful political figure in the world was using his Twitter account to undermine the bedrock of American democracy. And it was up to Roth to try to stop him.

"He'd violated the policy in exactly the type of way we had predicted," Roth said. "It was very direct, it was very straightforward. Either this was going to be our policy, or it wasn't. We were either going to rip the Band-Aid off and do it, or we weren't."

Roth ripped the Band-Aid off, putting a label on Trump's tweet that directed people to "get the facts about mail-in ballots." He explained Twitter's new approach to labeling misinformation in a corporate blog post. The young policy leader was trying to be transparent, but he was naive about how the move would be interpreted. The next day, Trump's senior adviser Kellyanne Conway was on Fox News calling Roth out by name. Right-wing pundits dredged up his old tweets, including one from 2017 where he'd asked "How does a personality-free bag of farts like Mitch McConnell actually win elections?" Overnight, Roth had become the face of censorship on Twitter, supposed proof that the platform had a left-wing bias.

These claims succeeded in making Twitter executives hesitant to take further action against Trump and his allies. Dorsey in particular hated bans. As the election approached and Trump's rhetoric became more inflammatory, the trust and safety team tried to roll out a new policy to cover coded incitements of violence, but executives chose not to enforce it, according to a deposition by Anika Collier Navaroli, a senior content policy expert at Twitter.

"Twitter saw President Trump's potential violent incitement of his supporters as a cause for concern even prior to Election Day but chose not to take effective actions to prevent him from using the platform in this way," read a scathing draft of a congressional report about social media's role in the leadup to the January 6 insurrection. "Moreover, this failure to act was consistent with Twitter's longstanding deferential treatment of President Trump."

Later that year, Roth began reporting to Vijaya Gadde, Twitter's chief legal officer, who would eventually become a target of Elon Musk's. At the time, she was known as a fierce free speech advocate. When Turkey banned Twitter for refusing to comply with a court order (reports of government corruption were circulating on the platform; the government wanted them taken down), Twitter filed two court petitions and successfully overturned the ban.

If Gadde's relationship with Jack Dorsey had stayed intact, none of this might have mattered. But it didn't. In 2021, Nigeria banned Twitter after the company deleted a tweet from President Muhammadu Buhari that threatened violence against separatists. Dorsey blamed Gadde for the ban, according to employees who worked closely with her. The following year, when Musk identified her as a censor, Dorsey was nowhere to be seen.

In 2020, Twitter's traffic soared as people stayed home glued to their screens. Monetizable daily active users—the number of people that Twitter could show ads to—were up 27 percent year over year. Dorsey went on a hiring spree, growing the company 53 percent between 2019 and 2021.

Yao Yue, one of Twitter's highest-ranking engineers, did not attribute Twitter's newfound success to Dorsey. In her mind, if Twitter was doing well, it was in spite of the CEO, not because of him.

Yue, a five-foot-three mother of two, was employee No. 300 at Twitter. She had a youthful face and a wry sense of humor that belied her technical chops. When she joined the company in 2010, Twitter did not have engineering levels, a standard structure elsewhere. Then, a few months after she started, the company imposed a ranking system. Yue went from being an engineer to being a "SWE 2." As employees discussed the new levels, Yue realized she was underpaid compared with some of her colleagues. She was building Twitter's caching system, a critical part of the engineering infrastructure that had to do with how information was stored and directly impacted the speed and efficiency with which Twitter could show users the information they wanted to see. In 2011, Yue went to her manager and asked for a raise. She'd come prepared to argue for what she wanted but found she didn't have to. Before she could even finish talking, her manager agreed. He said he was too busy to proactively check people's pay and thought that what she was suggesting was more than fair. He thanked her for bringing this to his attention. Yue even got a stock bump she hadn't asked for.

It was textbook Twitter: slow, bureaucratic, but quick to reward those who spoke up. Especially if those people were engineers.

Over the years, Yue learned to love Twitter's culture, which she described as "benevolent anarchy." She wasn't naive about the company's problems—Twitter was inefficient, its promotion process highly political—but she appreciated that anyone could influence the direction of the product, so long as they knew who to talk to and were willing to have many, many conversations.

Yue didn't believe in the gospel of hustle culture. In engineering, there were two questions to ask: Is this the right answer? And, is there a better way to solve this problem? Once she'd answered those queries, she didn't need to spend a million hours working just to prove that she cared about her job.

But some work was required, particularly from Twitter's leadership. As Dorsey continued to jet-set around the globe, seeming more checked out than ever before, a third question began floating in her mind: *How much longer is he going to remain CEO?*

CHAPTER 3

"Little Tentacles"

Randall Lin could barely sit still. It was February 2020, and he'd recently been hired as a machine-learning engineer at Twitter, which was far and away his favorite app. Already, the wiry twenty-nine-year-old was overflowing with ideas about how to make the recommendation algorithm better.

What Lin craved most of all was to get access to GPUs, supercomputing platforms that can process data very quickly. Twitter bought access to GPUs through Google Cloud, but this was like leasing a car instead of buying it and customizing it to one's specifications. Lin wanted to tinker, to optimize. Renting the GPUs from Google wasn't enough. He wanted to own the servers outright.

Unfortunately, Lin's colleagues didn't seem motivated to overhaul Twitter's machine-learning framework. The company had already spent a year building GPU servers before abandoning the project to move the work to the cloud. "People would either not do things, like not plan things, or they would spend all their time planning and never do things," Lin says.

Lin didn't want to step on anyone's toes. So he went to his engineering manager for advice. "I'm running into all these organizational-type problems," he said, attempting to be tactful. "What's your take on running an efficient org?"

The leader sent him a presentation outlining his management philosophy. It modeled Twitter's org chart after slime mold—an organism that sends out little tentacles to look for food, then gradually moves in the direction of the tentacle that finds the most sustenance. *Oooooh*, Lin thought when he saw the slide. *This is exactly what I'm feeling.*

Lin thought slime mold was a ridiculous management philosophy. It was the antithesis of how he liked to operate. Speed was his superpower. He thought of himself like an agile little mouse, not a slow protoplasm. "A mouse has to quickly explore and decide where to go to find food or hide," he explains. "But I understood that, at Twitter, I was one of those little tentacles, and I just had to go really deep and then shout, 'Come over here and look what I have! Don't you want this resource, too?' And then maybe the team would follow." Lin could be a good tentacle—he could make the rest of the slime mold pay attention.

Lin and his colleagues retrained Twitter's machine-learning models on the rented GPUs. He still believed he'd get better results if Twitter got its own machines, but he'd reluctantly put that dream on ice. Nevertheless, his results were good—users started spending more time on Twitter. After one of his experiments, the For You page saw a 3.5 percent increase in user active minutes, an almost unheard-of jump for a big tech platform of Twitter's size. During his first year, Lin was promoted to staff engineer, a trajectory that typically took two, if not three, years.

Still, the pointless bureaucracy rankled. Lin couldn't understand why his colleagues just accepted that, under Jack Dorsey, Twitter moved like molasses. Sure, it was convenient that tweeps could work from anywhere. But did the office need to be basically empty? And the lack of urgency felt wrong.

Lin had a strong sense of purpose, especially when it came to engineering. If he had the right answer, if he knew the solution to a problem, he didn't want to wait for another team to test various solutions. He wanted to *go*.

Though he chafed at the laxity of Twitter's culture, he wasn't an absolutist. As a physics student at the California Institute of Technology in

Pasadena, Lin had served on the Board of Control, a student committee that investigated allegations of honor code violations.

The honor code was a big deal at Caltech. Most exams were open book. Professors trusted students to have the integrity not to cheat.

Lin still remembered the international student who was accused of copying answers on an applied mathematics exam, a notoriously difficult class that all engineering majors had to take.

When the case came before the board, the committee reviewed the evidence and debated what to do. The investigation stretched on for almost twelve hours. The international student denied any wrongdoing, but some of her explanations were evasive. Could they trust her? Then one of the committee members noticed that, on one of the student's exam answers, she'd erased a variable, the imprint of which could still be seen on the paper. The variable didn't make sense. It wasn't part of the answer in question. On a whim, the committee looked online and saw that the variable matched an answer sheet that someone had posted for the test. The committee ruled that the student had likely copied the incorrect answers, then noticed her mistake and erased the evidence. They told the school that she should be suspended.

In every other case that Lin had been a part of, the school had always taken the board's suggestion. But this time, school administrators pushed back, saying that because the student wasn't from the United States, it was hard to know what kind of cultural norms and pressures she was up against, according to Lin's recollection. The school gave the student a lighter punishment. The committee was sworn to secrecy.

At first, Lin was incensed. He hated that the student was getting special treatment because she wasn't from the United States. But over time, he began to see the incident in a new light. "Rules are in place to encourage certain behaviors in a system," he said. "But mercy is how you stop people from being caught in a system. The trick is to make sure that mercy doesn't erode the rules."

Lin understood that Twitter needed to have rules. But he couldn't fathom why employees insisted on following the stupid ones, rather than thinking for themselves. It was like the whole company had taken on Dorsey's mindset: don't worry about shipping new features, don't come into the office, don't fret about increasing the share price. It was particularly frustrating because, as a user, Lin *loved* Twitter. He posted on a pseudonymous account that had amassed a few thousand followers by shitposting about artificial intelligence. During the pandemic, when everyone was stuck at home, he'd made real-life friends through the app. He understood how incredible Twitter could be. If only Dorsey would get out of the way.

CHAPTER 4

"Be There, Will Be Wild!"

In November 2020, Donald Trump lost the US presidential election to Joe Biden, but refused to accept the results. In the weeks that followed, he repeatedly claimed that the election had been stolen. On December 19, he told his followers there would be a "big protest" in Washington, DC, on January 6, 2021, the day that Congress was scheduled to ratify the electoral college votes and confirm Biden's victory. "Be there, will be wild!" Trump tweeted. A small footnote on the tweet warned that "this claim about election fraud is disputed."

Twitter's trust and safety team warned that Trump's online rhetoric would likely lead to real-world harm. As Roth's colleague Anika Collier Navaroli later testified in front of Congress, the existing coded incitement policy was unlikely to "lower the temperature on the platform." But they were not aligned about what to do about it.

On January 6, 2021, following a Trump rally on the Ellipse, thousands of pro-Trump rioters stormed the United States Capitol. Seven people died, hundreds more were injured. The damages exceeded $2.7 million. One former Twitter employee told congressional investigators that "the attack would not have occurred with the same magnitude" were it not for Trump's tweet.

Across the company, employees were furious that leadership had ignored their warnings. They called on Dorsey to permanently suspend Trump's account. "For the last four years, we have watched right wing extremists grow on our platform, nurtured by @realDonaldTrump," they wrote in an internal letter, which was signed by more than three hundred employees. "We have seen Twitter leadership struggle to deal with the violent, hateful rhetoric shared by @realDonaldTrump . . . we request that @realDonaldTrump's account be suspended permanently, before he can further harm using our platform."

Yoel Roth and Del Harvey, then the head of trust and safety, recommended that the company ban Trump. Dorsey overruled the decision and instead put Trump in time-out. Twelve hours later, Trump's account was restored, and he soon began praising his supporters, a violation of Twitter's "Glorification of Violence" policy. Dorsey finally reversed his stance. On January 8, Twitter permanently suspended the former president. Dorsey was working on a private island in French Polynesia when he made the call. "Trump's suspension ended the preferential treatment Twitter gave his account for years," the congressional committee wrote.

If banning Trump from Twitter was the right decision, critics could argue, convincingly, that it was one made too late. In conservative circles, the narrative about election fraud had taken root, as had the notion that Twitter employees like Gadde and Roth were abusing their power by pushing a liberal agenda. No matter that Twitter had allowed Trump to flout the rules for years. If tech companies could silence the most powerful man in the world, what did that mean for everyone else?

CHAPTER 5

"Woke Mind Virus"

Twitter's decision to ban Donald Trump was a turning point for Elon Musk. He loved the platform but he felt like it was sliding in a bad direction, gripped by a "woke mind virus" that was anti-meritocratic and anti–free speech. Never mind that the evidence showed Twitter actually amplified conservative voices. Once, a Twitter executive had called the company "the free speech wing of the free speech party." Now, all Musk saw was censorship.

Musk did not respect Trump, but he didn't think the former president should be banned from Twitter. Perhaps, on some level, he appreciated how Trump wielded his platform. If there was anything Musk admired outside of engineering talent, it was someone who knew how to tweet. Plus, there was undeniable synergy in the men's views on censorship. Musk had long been friendly with Jack Dorsey. But that wouldn't stop him from criticizing Twitter's overreach.

For Musk, the issue was becoming personal. In 2020, Musk's eldest child, Jenna, who is trans, stopped speaking to him, according to his biographer, Walter Isaacson. She'd become a Marxist and disavowed all her father stood for. Two years later, she successfully petitioned the court for a name change, first and last. In the petition, she explained the factors behind the decision as "Gender Identity and the fact that I no longer live with or wish to be related to my biological father in any way, shape or form."

Musk blamed the rift on Jenna's elite Los Angeles private school, Crossroads, which he claimed had infected his child with the "woke mind virus."

It wasn't just about pronouns. The future of humanity was at stake. "Unless the woke mind virus, which is fundamentally antiscience, antimerit, and antihuman in general, is stopped, civilization will never become multiplanetary," Musk told Isaacson. Incidentally, it was making Twitter less funny.

On November 29, 2021, ten months after Twitter banned Trump, Dorsey abruptly announced he was stepping down as CEO, effective immediately. He was unhappy with what Twitter had become and he lacked the power to reshape it. Unlike, say, Mark Zuckerberg, who'd retained near-total control of Meta through a dual-class share system that gave him more voting power than regular shareholders, Dorsey owned the same type of stock as everyone else.

Dorsey had made a few halfhearted attempts to change Twitter. For one, he had attempted to install his friend Elon Musk on the board of directors. But the board rejected the proposal, feeling that Musk was too risky, according to text messages between Dorsey and Musk that were later revealed in court as part of a lawsuit brought by Twitter. Dorsey thought the move was "completely stupid and backwards."

"That's about the time I decided I needed to work to leave, as hard as it was for me," he said.

By the time he resigned, Dorsey no longer believed that Twitter should even be a company.

"I believe it must be an open source protocol, funded by a foundation of sorts that doesn't own the protocol, only advances it," he texted Musk in March 2022. "It can't have an advertising model. Otherwise you have a surface area that governments and advertisers will try to influence and control. If it has a centralized entity behind it, it will be attacked."

Musk was one of the only people wealthy and reckless enough to realize Dorsey's vision. In January, he had begun buying up Twitter shares, quietly accumulating a massive stake in the company.

Ever since he was a child, Musk had harbored a belief that he was destined to have a great impact on the world. While he'd originally wanted to design video games, he ultimately chose to become an entrepreneur because game design didn't seem grand enough. "I really like computer games, but then if I made really great computer games, how much effect would that have on the world," he said. "It wouldn't have a big effect . . . I couldn't bring myself to do that as a career."

"He started to get a sense that if he didn't exist, the world would die," says Kara Swisher. "He told me, essentially, that if his companies didn't make it, humanity was fucked."

Through Tesla, he was pushing the world toward clean energy. Through SpaceX, he was ensuring the human race's survival even if Tesla failed to save Earth. Now, through Twitter, he believed he had a chance to save democracy and restore free speech to the digital town square. It didn't hurt that Musk saw Twitter as an "accelerant to creating X, the everything app," the social payments platform that PayPal could've been.

Walter Isaacson, Musk's biographer, thought this was a bad idea. "This is not well suited for him," Isaacson said during a Twitter Spaces interview in 2023. "He doesn't have a feel for human emotion, and he's got incredibly strong opinions, which makes him somebody who is good at tweeting but doesn't make you good at running Twitter if you're trying to make it a level playing field."

Musk said he wanted less drama in his life, but he couldn't get Twitter off his mind. He dreamed of restoring the platform to a free speech haven while simultaneously turbocharging its potential.

Anyone who worked at Twitter could have told him these goals were in direct conflict. Users might disagree about the type of content moderation they wanted to see, but few wanted no moderation whatsoever, which could devolve into a harassment-laden hellscape. More importantly, Twitter was hamstrung by outside forces. As long as Twitter was a company—even a private company—it was beholden to app stores, regulators, and (in all likelihood) advertisers.

Yoel Roth's trust and safety team knew this intimately. For years, they'd been working to balance content moderation with free speech. If Twitter failed to enforce its policies, it would get a call from Apple, hinting that the company's next app release might be held up if it didn't take stronger action.

But Musk wasn't talking to the trust and safety team. He was talking to his peers, ultrawealthy tech bro libertarians, many of whom were out of touch with the experience of Twitter employees and regular users.

Court filings would later reveal that in March 2022 Musk was telling his friend and longtime investor in his companies, Valor Equity Partners founder and CEO Antonio Gracias, that free speech mattered most when it was "someone you hate spouting what you think is bullshit."

"I am 100% with you Elon," Gracias responded. "To the fucking mattresses no matter what this is a principle we need to fucking defend with our lives or we are lost to the darkness."

Musk reacted to the text with a heart.

SEC rules require investors to disclose their position within ten days of acquiring a 5 percent stake in a company. After quietly gobbling up shares of Twitter in early 2022, Musk hit that threshold on March 14, 2022, but he didn't file the necessary paperwork. He had little time for regulators, least of all those who worked for the SEC.

Still, Musk couldn't help hinting about what he was up to. "Free speech is essential to a functioning democracy," he tweeted on March 25, in a post that was accompanied by an interactive poll. "Do you believe Twitter rigorously adheres to this principle? The consequences of this poll will be important. Please vote carefully." More than two million accounts voted, with 70 percent saying no.

Musk's free speech poll piqued the interest of William MacAskill, a Scottish philosopher and architect of the effective altruism movement, which takes a utilitarian approach to philanthropy in an effort to maximize the benefit of every donation.

"I'm not sure what's on your mind, but my collaborator Sam Bankman-Fried has for a while been potentially interested in purchasing [Twitter] and then making it better for the world," MacAskill said, referencing the billionaire wunderkind behind FTX, then the world's second-biggest cryptocurrency exchange.

"Does he have huge amounts of money?" Musk asked.

"Depends on how you define 'huge'!" MacAskill said. "He's worth $24B, and his early employees (with shared values) bump that to $30B. I asked about how much he could in principle contribute and he said '~$1–3b would be easy ~$3–8b I could do ~$8–15b is maybe possible but would require financing.'"

Musk wasn't convinced. Since he'd started dropping hints about his plans for Twitter, his phone had been blowing up with messages from associates eager to share their two cents. He asked MacAskill if he vouched for Bankman-Fried.

"Very much so!" MacAskill responded. "Very dedicated to making long-term future of humanity go well."

Musk agreed to meet with the crypto kid, and the pair ended up talking for half an hour. But Musk felt like something was off. He often led with his intuition, and his intuition was telling him not to trust Bankman-Fried, no matter how many people thought he was a savant. "My bullshit meter was redlining," he said after FTX collapsed in November 2022. "There's something wrong, and he does not have capital, and he will not come through. That was my prediction." Likewise, Bankman-Fried thought Musk was nuts, according to Walter Isaacson.

In December 2022, Bankman-Fried was charged with seven criminal counts for allegedly defrauding investors, including securities fraud, money laundering, and campaign finance violations. He was eventually found guilty on all counts.

CHAPTER 6

"Welcome Elon!"

When Jack Dorsey stepped down on November 29, 2021, Twitter's board appointed chief technology officer Parag Agrawal to take his place. "He leads with his heart and soul and is someone I learn from daily," the founder said in an internal email announcing the change. "My trust in him as our CEO is bone deep."

Agrawal's appointment caught employees by surprise. At thirty-eight years old, he didn't have experience running a company—not to mention a public tech company of Twitter's size. "He was very, very naive, and completely overwhelmed," said Yao Yue, who briefly reported to him before he became CEO.

Still, there was no question that Agrawal was talented. At the age of seventeen, the Mumbai-born executive won a gold medal at the International Physics Olympiad in Antalya, Turkey, before going on to study computer science and engineering at the elite Indian Institute of Technology and pursue a PhD at Stanford University. In 2011, he joined Twitter as a software engineer, his first full-time tech job.

The even-keeled engineer helped Dorsey overhaul Twitter's infrastructure. Then he laid the groundwork for Twitter to become a decentralized protocol, a long-standing obsession of Dorsey's. The goal was to free up users to move between social platforms, carrying their tweets and followers with them wherever they went. This also would ease the burden on Twitter to moderate content. The platform was dealing with contradictory regulatory pressures—liberals wanting it to remove more content and conservatives wanting it to remove less. Decentralization offered a way out by letting people build their own nodes in the network and moderating them as they saw fit. The company called it Project Bluesky. In 2022, the team announced the formation of a public benefit LLC, noting Bluesky had become an independent organization.

If Bluesky had any immediate impact at Twitter, it was in bringing Dorsey and Agrawal closer together. Dorsey had become an unabashed Bitcoin enthusiast. Agrawal was one of the highest-ranking Twitter employees who took his blockchain ideas seriously. "Parag was Jack's technical guy," said Yue. "If Jack had a question, he would go to Parag."

In 2017, Agrawal was promoted from distinguished engineer to chief technology officer. Shortly after stepping into his new role, Agrawal had a meeting with a group of product leaders, where he tried to give an inspiring speech about the need for them to come together as a team. One of the directors raised his hand. "Why should we be a team?" he asked provocatively. The attendees looked at Agrawal, expecting a rousing (if conventional) response about the benefits of collaboration and mutual trust. But the new CTO was speechless. He seemed to have no idea what to say.

During this time, Twitter's workforce was expanding, and the engineering team needed a system to track what people were working on. Under Agrawal's guidance, the organization created a "universal priority list" (UPL) to rank technical projects by their potential importance to the company.

The list quickly became a joke internally. Its size made it hopelessly unwieldy. It included every project that involved more than one team, regardless of what kind of project it was, what it affected, or how likely it was to succeed. When the UPL launched, it had roughly five hundred projects on it.

Soon, product managers were jockeying to have their projects ranked first on the UPL. The prevailing wisdom was "if your project isn't one of the first fifty, don't bother trying to move it forward," one employee recalls.

Despite the growing pains he experienced in the C-suite, when Agrawal was appointed CEO, executives were quick to frame the move as a positive. Agrawal was "rigorous and incredibly sharp," wrote Kayvon Beykpour, Twitter's general manager, in an internal email. "The momentum and trajectory that we have at Twitter is 🔥," he added. "Our strategy has never been better and our team is stronger than ever," said Michael Montano, Twitter's head of engineering.

Agrawal promised his new workforce that there was "no limit to what we can do together." But Twitter needed to cut costs.

Dorsey had hired too many people at the height of the pandemic, miscalculating how long the advertising boom was going to last. In 2021, revenue was roughly $5 billion and the company's costs were almost $5.6 billion. The situation wasn't dire—Twitter had very little debt for a company its size, and the cyclical ads market would bounce back—but it required Agrawal to make some tough calls. The CEO started planning for layoffs. By April, if all went according to plan, Twitter would shed about 16 percent of its workforce, roughly twelve hundred people in total.

In December 2021, days after Agrawal was named CEO, the global design and research team converged in the Bay Area for a three-day offsite event. The company put them up in four-star hotels and took them wine tasting in Napa Valley.

Dantley Davis, Twitter's chief design officer, cohosted a miniconference so people could hear about what their colleagues were working on. There was a talk about advocating for the user and a presentation about non-fungible tokens (NFTs).

Davis was a polarizing figure at Twitter. Four months earlier, *The New York Times* published an exposé about his blunt management style, which had prompted several HR investigations. Nikkia Reveillac, Twitter's beloved global head of research, had raised these concerns to Jack Dorsey before the article came out, only to be abruptly fired.

Reveillac's ouster rocked the team, which had been through a bruising couple of years. Researchers were responsible for studying the dynamics of hate speech, harassment, and misinformation. Every day seemed to bring a fresh scandal and scathing criticism from both conservatives and liberals.

Davis used the off-site event as an opportunity to rebuild trust in his leadership, according to one attendee. He laid out a vision for what the team could accomplish in 2022. He assured employees that Agrawal was supportive of their work and would give them the resources they needed to achieve their goals.

By the end of the retreat, the mood had shifted. The team felt unified and hopeful. Davis flew out early but employees met for a final goodbye brunch. As they ate pastries and made DIY flower arrangements, phones started dinging. A collective gasp was heard around the table, followed by a spontaneous burst of tears. Those who weren't looking at their phones were perplexed. "I was just sitting there eating a croissant being like, 'What's up with the mood?'" one researcher remembers.

Agrawal had laid off Dantley Davis and Michael Montano, the head of engineering, as he began to reshape the executive ranks.

The following month, Rinki Sethi, the company's chief information security officer, and Peiter "Mudge" Zatko, the head of security, were also cut. The company said Mudge was fired "for ineffective leadership and poor performance," but later agreed to pay the former security chief a $7 million settlement.

In February 2022, Agrawal went on a short parental leave. He wanted to show employees that work-life balance started at the top ("I told him to take paternity leave to set the right example," Yue said). While he was out, he continued talking to the executive team every day. But most employees didn't hear from him directly. Tweeps had been cautiously optimistic that he would be a strong leader after the executive layoffs, but fears about his leadership started to spread in lower levels of the company as employees worried he didn't have what it took to do the job. In trying to model a healthy approach to work, Agrawal unwittingly created the illusion of a power vacuum.

Parag Agrawal had big plans for his first year as Twitter's CEO. They did not include going to battle with the richest man in the world.

Four months into his tenure, however, it was clear he wouldn't have a choice. On April 4, 2022, eleven days after Elon Musk was supposed to disclose his position to the SEC, he finally revealed that he had a significant stake in Twitter. Not only that, but he owned 9.1 percent of the company. While Agrawal had been getting up to speed as CEO, Musk had become his largest shareholder.

Agrawal and Bret Taylor, co-CEO of Salesforce and chairman of Twitter's board of directors, tried to suss out Musk's intentions. The stake he'd acquired suggested he might try to take over the company. But the paperwork he'd filed with the SEC suggested otherwise.

If Musk was plotting a corporate coup (and complying with SEC guidelines), he would've filed a 13D, a detailed form meant for active investors. But he filed a 13G, a shorter form reserved for passive investors.

With Musk, it was hard to tell what was part of a strategy and what was simply sloppy paperwork. He was trying to get to Mars—he didn't have time to fill out forms. As if to underline this point, the original 13G Musk filed said he owned 9.2 percent of Twitter rather than the 9.1 percent stake he said he owned in subsequent filings. Even if the mistakes were purely accidental, they turned out to be useful. Musk had continued buying shares after the March 24 disclosure deadline. Once news about his position became public, Twitter's share price jumped from $39 to $50. By hiding his Twitter position for eleven days, he'd saved himself roughly $143 million, according to a Twitter shareholder lawsuit. This also meant that Twitter shareholders who'd sold stock during that eleven-day period did so at a lower price than they otherwise might have been able to.

On March 31, days before the news broke, Musk was in the Bay Area for a Tesla Autopilot meeting. Agrawal and Taylor, who'd just gotten wind of his position, asked Musk to meet. He agreed, and Taylor changed his flight from New York to get in early so the trio could speak over dinner. Taylor booked an Airbnb near the airport.

"This wins for the weirdest place I've had a meeting recently," he texted Musk when he arrived, sounding self-conscious, according to text messages later revealed in court.

"Haha awesome," Musk said gamely. "Maybe Airbnb's algorithm thinks you love tractors and donkeys (who doesn't!)."

Taylor kept the joke going. "And abandoned trucks in case we want to start a catering business after we meet," he said.

"Sounds like a post-apocalyptic movie set," Musk responded.

That night, Agrawal pitched Musk on joining Twitter's board of directors. Musk was open to the idea. He found Agrawal nice—but in his mind, this could be a liability. "What Twitter needs is a fire-breathing dragon, and Parag is not that," he said.

Musk couldn't help feeling like he had a better grasp of Twitter than either Agrawal or Taylor. Every time he opened the app, he was greeted by a steady stream of spam. Agrawal and Taylor barely tweeted. Perhaps they weren't aware of the problem. "Crypto spam on Twitter really needs to get crushed," Musk texted them on March 31, shortly before arriving at dinner. "It's a major blight on the user experience and they scam so many innocent people." (According to the text messages, neither Agrawal nor Taylor responded to this piece of advice.)

On April 4, Musk's stake in Twitter became public, sending its stock soaring 27 percent. It was a good sign for Twitter's board, and even better for Elon Musk, who stood to make over $700 million on his investment.

Later that day, Musk polled his followers to ask if they wanted Twitter to roll out an edit button—a feature request that had become so ubiquitous it was a meme. Seventy-three point six percent of respondents said yes.

Within twenty-four hours, Twitter announced: the edit button was coming soon.

"now that everyone is asking . . . yes, we've been working on an edit feature since last year! no, we didn't get the idea from a poll 😉," the company wrote.

Musk's friends didn't believe it for a second. "And you already got them to try the edit!" texted Ira Ehrenpreis, a venture capitalist who sat on Tesla's board of directors. "Oh yeah . . . it had been in the works. Sure."

Around this time, Yoel Roth recalls that he got a surprising request regarding Elon Musk. Leslie Berland, Twitter's chief marketing officer, asked him to transfer ownership of the @e handle to Twitter's new potential board member.

On Twitter, single letter usernames were a hot commodity. They signified status—either the user was an early adopter, a good hacker, or simply had a lot of money. On black market forums a handle like @e could go for $50,000.

As a rule, Twitter rarely released high-value usernames unless there was a legitimate business association between the name and a particular user. There seemed to be no legitimate business claim between Musk's brand and the @e handle. Staffers asked Roth what they should do.

Roth investigated and found that the current account owner had gotten the username by stealing it from someone else. As a consequence, the account had been suspended. "Sure, give it to Elon," Roth said. Why not?

Even then, Musk wasn't a regular user, he was a power user, maybe even *the* power user. Musk got the account, changed the name to "John Utah," and never tweeted.

On April 4, 2022, Twitter sent Musk the paperwork for joining the board. To his surprise, the offer included a number of provisions barring him from speaking ill about the company. Musk found it ironic that Twitter was trying to restrict his freedom of speech. "Thank you for considering me for the Twitter board, but, after thinking it over, my current time commitments would prevent me from being an effective board member," he texted Bret Taylor. "This may change in the future."

Within hours, Twitter sent him a new agreement. This time, there were no speech restrictions. Musk accepted the board seat.

On April 5, Agrawal tweeted that Musk was joining Twitter's board of directors, in an announcement he let Musk approve. "Through conversations with Elon in recent weeks, it became clear to us that he would bring great value to our Board," he posted. "He's both a passionate believer and intense critic of the service which is exactly what we need on @Twitter, and in the boardroom, to make us stronger in the long-term. Welcome Elon!"

CHAPTER 7

"Midas Touch"

Twitter employees woke up on April 5, 2022, to see the news that Elon Musk—storied entrepreneur, burgeoning right-wing troll—was joining the company's board. Slack lit up as angry employees demanded answers.

"We know that he has caused harm to workers, the trans community, women, and others with less power in the world," one employee wrote on Slack. "How are we going to reconcile this decision with our values? Does innovation trump humanity?"

What could Agrawal say? Since early 2020, Musk had railed against the Covid lockdowns on Twitter, retweeted a Hitler meme, and replied "I keep forgetting that you're still alive" in a derisive tweet to Bernie Sanders. Agrawal could hardly expect his liberal Bay Area workforce to rejoice at the Tesla CEO's arrival.

Agrawal thought the best route forward was for Musk to speak to employees directly. He'd seen firsthand that the billionaire could be far more reasonable face-to-face than the provocative character he played on Twitter.

"I expect most questions to not get into specific ideas/depth—but more around what you believe about the future of Twitter and why it matters," he told Musk in a text message. He warned him that there would likely be questions "from people who are upset that you are involved and generally don't like you for some reason." The goal, he said, was "for people to just hear you speak directly instead of making assumptions about you from media stories."

Musk agreed, and Agrawal sent out an email announcing the meeting. "Following our board announcement, many of you have had different types of questions *about* Elon Musk, and I want to welcome you to ask those questions *to* him."

Employees were not pleased. The email promptly leaked to *The Washington Post* along with scores of Slack messages highlighting their concerns. An employee who said he'd worked at Tesla claimed he'd "witnessed the awful changes in company culture that followed" when Musk became CEO. "I'm extremely unnerved right now, because I've seen what he can do firsthand," the employee wrote on Slack.

Agrawal scrambled to reassure his incoming board member. He texted Musk a link to the article, adding, "I think there is a large silent majority that is excited about you being on the board, so this isn't representative. Happy to talk about it—none of this is a surprise."

The meeting was eventually scrapped, but for the moment Musk's response was conciliatory. "As expected," he said. "Yeah, would be good to sync up. I can talk tomorrow night or anytime this weekend. I love our conversations!"

Musk had a ton of ideas for how to make Twitter better. "[Let me know] if I'm pushing too hard," he added, as if testing out how far the CEO would go to keep him happy. "I just want Twitter to be maximum amazing."

"I want to hear all the ideas—and I'll tell you which ones I'll make progress on vs. not. And why," Agrawal responded. His tone was appeasing, but he seemed to be trying to remind Elon who had the power to make those ideas real.

One of Musk's ideas was to reshape Twitter's leadership team. To that end, Musk began pressuring Agrawal to fire Vijaya Gadde, who he saw as Twitter's "top censorship advocate," perhaps due to her role in the decision to ban Donald Trump. Agrawal wouldn't hear it. Gadde had a sterling reputation at Twitter and had forcefully defended free speech around the world. Agrawal had no intention of letting her go.

As the relationship between Agrawal and Musk grew more strained, employees who didn't fit into Twitter's laid-back culture were rejoicing. "Elon Musk is a brilliant engineer and scientist, and he has a track record of having a Midas touch, when it comes to growing the companies he's helped lead," wrote Luke Simon, a senior engineering director, on Slack.

Like Randall Lin, Simon bristled at the company's relaxed atmosphere, which had contributed to the notorious paralysis in its slow-to-ship product organization. He told his team that he wanted to build an "impact focused, egalitarian and empirical culture, where any team member, with a strong data-driven justification, gets the metaphorical center stage." Simon dreamed of a more swashbuckling version of Twitter, in which he played a starring role—coworkers had seen a photo of him posing in front of an oil painting of himself dressed as Napoleon.

Some of his colleagues disagreed. "I take your point, but as a childhood Greek mythology nerd, I feel it is important to point out that the story behind the idea of the Midas touch is not a positive one," one responded. "It's a cautionary tale about what is lost when you only focus on wealth."

CHAPTER 8

"Back Door Man"

Musk's friends were thrilled to hear that one of their own was infiltrating Twitter.

"Excited to see the stake in Twitter—awesome," texted Joe Lonsdale, managing partner of the venture capital firm 8VC, on April 4. "'Back door man' they are saying haha. Hope you're able to influence it. I bet you the board doesn't even get full reporting or see any report of the censorship decisions and little cabals going on there but they should—the lefties on the board likely want plausible deniability!"

Joe Rogan, a wildly popular podcast host, texted Musk to ask if he planned to "liberate Twitter from the censorship happy mob."

"I will provide advice, which they may or may not choose to follow," Musk responded magnanimously.

But his mind was already spinning with possibilities. What if, instead of joining Twitter's board, he built a competitor, resurrecting X, the "everything app" he'd tried to start at PayPal? What if he put it on the blockchain? "I have an idea for a blockchain social media system that does both payments and short text messages/links like Twitter," he texted his brother, Kimbal Musk, on April 8, before laying out a convoluted pay-to-play scheme for a blockchain-based social network.

Musk described a "massive real-time database that keeps a copy of all blockchain messages in memory, as well as all messages sent to or received by you, your followers and those you follow" culminating in a "twitter-like app on your phone that accessed the database in the cloud."

"This could be massive," he said.

It sounded like the musings of a stoned fifteen-year-old who'd watched one too many blockchain dudes on YouTube. But his brother was encouraging, perhaps hopeful that it would keep Elon from getting too wrapped up in Twitter. "I'd love to learn more," Kimbal told him.

Kimbal Musk had worked with Elon since the '90s, when he'd helped his brother start Zip2. He was one of the few people who was willing to give his brother honest feedback. And his feedback about joining Twitter's board was: *don't do it*, according to Walter Isaacson. "You don't know how much this is going to suck for you," Kimbal told his brother. "You tell people what you think, and then they smile and nod and ignore you." Creating an alternative—particularly a technically complex alternative on the blockchain, then one of the buzziest technologies—seemed like a better use of Elon's skills.

Musk told his lieutenant Steve Davis about his plan B if the Twitter board seat didn't work out. He'd build a "blockchain-based version of twitter" where users paid a small amount of Dogecoin to tweet. Rather than rid Twitter of crypto spam, he'd build his own app and integrate crypto into the very foundation of the product. Like Kimbal, Davis, a former bar-owner-turned-CEO of Musk's tunnel-digging firm, The Boring Company, seemed to take the half-baked idea seriously. "Amazing!" he said.

It took Musk a little over two weeks to realize that, for technical reasons, his dream of "blockchain Twitter [wasn't] possible." It would take too much computing power to ferry user data around the globe. In the end, only big entities would be able to handle it, which defeated the purpose of running a decentralized network.

Meanwhile, another idea was starting to take shape in Musk's mind. A vision where he didn't sit on Twitter's board—he sat on the throne, owning the company outright. Musk was growing frustrated with Twitter's management. "They're nice, but none of them use Twitter," he told his brother, Kimbal. "I don't feel anything will happen." He knew buying Twitter was a big risk on multiple levels. But if Musk had a superpower, it was that he'd always liked taking risks.

On April 9, 2022, Musk decided it was time to stop playing nice with Parag Agrawal. He'd noticed that some high-profile users, including Taylor Swift and Justin Bieber, had stopped tweeting. "Is Twitter dying?" he asked his followers.

Twitter's marketing team was furious. They'd been working for months to get Swift back onto the platform, despite the torrent of abuse she experienced every time she opened the app. Musk's tweet wasn't helping.

Agrawal had no choice but to intervene. "You are free to tweet 'is Twitter dying?' or anything else about Twitter—but it's my responsibility to tell you that it's not helping me make Twitter better in the current context," he wrote in a text message later revealed in court. "Next time we speak, I'd like to provide you perspective on the level of the internal distraction right now and how it's hurting our ability to do work . . . I'd like the company to get to a place where we are more resilient and don't get distracted but we aren't there right now."

Musk could barely hide his contempt. "What work did you get done this week?" he asked. Then he confirmed the executive's fear. "I'm not joining the board," he said. "This is a waste of time."

Then: "Will make an offer to take Twitter private."

"Can we talk?" Agrawal responded frantically, moments later.

Musk didn't answer, so Agrawal called Bret Taylor. Taylor also tried getting Musk on the phone.

"Parag just called me and mentioned your text conversation," he said. "Can you talk?"

"Please expect a take private offer," Musk responded tersely.

"I saw the text thread," Taylor said. "Do you have five minutes so I can understand the context? I don't currently."

"Fixing twitter by chatting with Parag won't work," Musk said. "Drastic action is needed."

The following week, Musk learned that Agrawal was on a family vacation. *Unbelievable.*

"Btw, Parag is still on a ten day vacation in Hawaii," he texted his buddy Jason Calacanis, an angel investor who cohosted the popular tech podcast *All-In.*

"No reason to cut it short . . . in your first tour as ceo," Calacanis joked. "Shouldn't he be in a war room right now?!?"

"Does doing occasional zoom calls while drinking fruity cocktails at the Four Seasons count?" Musk asked.

Agrawal had landed on Musk's bad side by refusing to fire Vijaya Gadde. Now he'd made the unforgivable decision to take a vacation. Musk had once joked that taking a trip to South Africa with his then wife, Justine Musk, had nearly killed him (he'd contracted a bad case of malaria on the trip). "I came very close to dying," he said. "That's my lesson for taking a vacation: vacations will kill you."

CHAPTER 9

"I Made an Offer"

By the time Musk's jet touched down in Vancouver, British Columbia, on April 13, 2022, the Tesla CEO was manic. He'd been pounding Red Bull for hours, turning the Twitter decision over in his mind. Ostensibly, the reason for the trip was personal. Musk's on-again, off-again girlfriend, Claire Boucher, the pop star known as Grimes, wanted to introduce their son, the unfortunately named X Æ A-Xii, to her parents and grandparents who lived outside Vancouver. But Musk rarely had only one reason for doing anything. The last day of the TED Conference was April 14. Musk was scheduled to be the keynote speaker.

Musk and Boucher were supposed to go see her family on April 13, but Musk was too wired, so he stayed back at the hotel. That afternoon, he texted Bret Taylor to let him know he was going to make an offer to take Twitter private.

"I am offering to buy 100% of Twitter for $54.20 per share in cash, a 54% premium over the day before I began investing in Twitter and a 38% premium over the day before my investment was publicly announced," Musk wrote in the offer letter. "My offer is my best and final offer and if it is not accepted, I would need to reconsider my position as a shareholder."

Musk continued, explaining why he felt the pivot from board member to potential owner was warranted—even necessary.

"I invested in Twitter as I believe in its potential to be the platform for free speech around the globe, and I believe free speech is a societal imperative for a functioning democracy," he wrote. "However, since making my investment I now realize the company will neither thrive nor serve this societal imperative in its current form. Twitter needs to be transformed as a private company."

That night, Musk stayed awake playing *Elden Ring*. Like many fantasy role-playing games, it cast the player as a powerful hero, vanquishing giant monsters with swords and spells. Video games helped Musk shut out other distractions, allowing him to process his thoughts in peace. Early the next morning, he sent a tweet: "I made an offer." Finally, his vision for Twitter was out in the open.

Initially, analysts scoffed. There was little evidence to suggest Musk's offer to buy Twitter at $54.20 a share was based on a detailed financial model. The pricing was meant to shoehorn in a weed joke. (The same joke that had previously gotten him in trouble with the SEC.) Twitter shares had been trading at $68 just six months earlier and hit an all-time high in March 2021. "No one believes this is the final price. No board in America is going to take that number," said Jefferies stock analyst Brent Thill in an interview with *Yahoo Finance*.

Saudi Arabia's Prince Alwaleed bin Talal, who was a major Twitter investor, flatly rejected Musk's bid, tweeting that it didn't "come close to the intrinsic value" of the company.

Even if Twitter's board of directors agreed, they still needed to take Musk's offer seriously. On April 15, they unanimously adopted a limited duration shareholder rights plan, more commonly known as a "poison pill," to stop Musk from completing a hostile takeover. Under the plan, if Musk purchased 15 percent or more of Twitter's outstanding stock without board approval, the company's remaining shareholders would be able to buy additional shares at a discount, diluting his stake in the company. The board had a fiduciary duty to seriously consider Musk's offer, but it wanted to explore its options.

Hours after the bid became public, Musk walked onstage in front of eighteen hundred people and sat down in a chair across from Chris Anderson, head of the nonprofit media organization TED. Anderson hadn't been sure Musk would make it to the conference. One week earlier, the British American businessman had recorded an interview with Musk at the Tesla Gigafactory in Austin, Texas, in case Musk didn't show. Now, here was Anderson, talking to the visionary onstage.

"So, Elon, a few hours ago, you made an offer to buy Twitter," Anderson said, enunciating each word as if he still couldn't believe the news himself. "Why?"

Musk raised his eyebrows at the audience, then turned away shyly. "I don't know," he said. "By the way, have you seen the movie *Ted* about the bear?"

"I have, I have," Anderson responded politely.

"It's a good movie!" Musk said, letting out a nasally laugh. Then he appeared to remember why he was there. "Was there a question?"

"Why make that offer?" Anderson prompted.

"Well, I think it's very important for there to be an inclusive arena for free speech . . ." Musk said. A person in the audience let out a solitary "whoop!"

"Twitter has become the de facto town square," Musk said, as if he was working through his rationale in real time. "It's just really important that people have the reality and the perception that they're able to speak freely within the bounds of the law, and so one of the things that I believe Twitter should do is open source the algorithm and make any changes to people's tweets, if they're emphasized or deemphasized, that action should be made apparent so anyone can see that action has been taken, so there's no behind-the-scenes manipulation, either algorithmically or manually."

Anderson pressed him. "But last week when we spoke, Elon, I asked whether you were thinking of [doing this], and you said, 'No way.' You said, 'I do not want to own Twitter, it is a recipe for misery, everyone will blame me for everything.' What on earth changed?"

"No, I think everyone will still blame me for everything," Musk said. "If I acquire Twitter and something goes wrong, it's my fault, 100 percent. I think there will be quite a few errors."

Later in the interview, Anderson asked Musk a question that harked back to his infamous 2018 tweet about taking Tesla private. "Is funding secured?" he asked.

The audience laughed.

"I have sufficient assets," Musk said, stroking his chin. It wasn't an answer.

CHAPTER 10

"I'd Jump on a Grande for You"

Now that Elon Musk had offered to buy Twitter for $44 billion, he needed to come up with the money. He might've been the richest man in the world, with a net worth that hovered around $219 billion, but that wealth was relatively inaccessible, as much of it was tied up in Tesla stock.

"Any interest in participating in the Twitter deal?" he asked his friend Larry Ellison, the billionaire CEO of Oracle, who sat on Tesla's board and owned $1 billion of Tesla stock, on April 20.

"Yes, of course," Ellison responded. He suggested he could offer "a billion . . . or whatever you recommend."

Musk suggested $2 billion or more. "This has very high potential and I'd rather have you than anyone else," he added.

Musk wanted the deal to seem exclusive. This was how business got done at the upper echelons of Silicon Valley. He offered friends the "opportunity" to invest. He didn't just ask people for money.

Unfortunately, not all of Musk's friends got the memo about the classy way to pitch. Jason Calacanis, the angel investor and podcast host, blasted the news out to his network. "We are now collecting interest to invest in Twitter with Elon Musk's plan to take it private," he said. The minimum investment in Calacanis's fund was $250,000—a far cry from the $2 billion Musk suggested to Ellison.

Musk was furious. "What's going on with you marketing [a special purpose vehicle] to randos?" he asked, according to text messages later revealed in court. "This is not ok."

"Not randos," Calacanis said. "I have the largest angel syndicate and that's how I invest. We've done 250+ deals like this and we know all the folks. I thought that was how folks were doing it . . . just wanted to support the effort." Later he added: "There is *massive* demand to support your effort btw . . . people really want to see you win."

Musk didn't seem to believe him. He'd talked it over with Jared Birchall, a wealth manager who ran his family office and acted as CEO of Musk's brain-implant company, Neuralink, and told Calacanis that Birchall was suspicious of his motives. "Morgan Stanley and Jared think you are using our friendship not in a good way," Musk texted the angel investor. "This makes it seem like I'm desperate. Please stop."

"Only ever want to support you," Calacanis replied, appearing chastised.

"Morgan Stanley and Jared are very upset."

"And you know I'm ride or die brother," Calacanis said. "I'd jump on a grande for you." (He presumably meant to write "grenade.")

Later, I reached out to Calacanis for comment, even offering to speak off the record. "The last thing I would ever do in my life is talk off the record about a friend," he said. He tried to cc my boss, Casey Newton, but mistyped his email address.

As Twitter's board negotiated the terms of the agreement with Musk, the markets took a turn for the worse. By the end of April, the S&P 500 had fallen 6 percent. Musk's offer, which had initially looked low, became more attractive to Twitter shareholders.

Twitter's banking advisers, including Goldman Sachs and J.P. Morgan, ran the numbers to see whether the company's stock could recover if the deal didn't go through. The results were grim. Twitter's board doubted Parag Agrawal could get Twitter's share price to $54.20 anytime soon.

On April 25, the board accepted Musk's proposal. Twitter hadn't made Musk come up in price ($54.20 was in fact his "best and final offer"), but it did secure a seller-friendly purchase agreement, including an unusually high reverse breakup fee of $1 billion.

Some commentators took this to mean Musk could get out of the deal if he paid $1 billion—but that interpretation was incorrect. The terms of the deal stated that Musk could only get out of the purchase agreement (and pay the breakup fee) if there was an outside reason the deal could not close. These reasons included if regulators blocked the agreement (unlikely), if the debt financing fell through (also unlikely, as Musk had already secured commitment letters from the banks), or if Twitter engaged in fraud. Even then, Musk would have to prove that the fraud had a "material adverse effect" on the business. Musk had gone as far as to propose deal terms that waived his right to due diligence, according to Twitter's proxy statement.

The agreement stipulated that Musk could continue tweeting about the merger so long as his tweets did not disparage the company or any of its representatives. Twitter executives speculated he hadn't read the final document.

Later that day, Twitter and Musk issued a joint press release announcing the agreement. Musk was taking Twitter private at $54.20 a share in a cash transaction that valued the company at about $44 billion—a 38 percent premium over Twitter's stock price before Musk's stake in the company was revealed.

"Free speech is the bedrock of a functioning democracy, and Twitter is the digital town square where matters vital to the future of humanity are debated," Musk said in a statement. "I also want to make Twitter better than ever by enhancing the product with new features, making the algorithms open source to increase trust, defeating the spam bots, and authenticating all humans. Twitter has tremendous potential—I look forward to working with the company and the community of users to unlock it."

Agrawal knew that he'd effectively signed his own death warrant. But what choice did he have? His contract included a "golden parachute" clause that stipulated that if Musk fired him after the deal closed, Agrawal would get roughly $57.4 million—an incentive meant to encourage him to put the good of the shareholders before his own career interests. Employee interests didn't seem to even be on the table. No one had fiduciary responsibilities to them.

On April 25, Agrawal hosted an all-hands meeting to try to address employees' concerns. One employee asked whether Musk would reinstate Donald Trump's account. "Once the deal closes, we don't know which direction the platform will go," Agrawal said opaquely.

The next day, Twitter published an acquisition FAQ to try to encourage employees to stick around once the deal closed. The document said their employment agreements wouldn't change for at least twelve months once Musk took over. Their stock would continue to vest and they'd be paid out on schedule. "The terms of the agreement specifically protect Tweep benefits, base salary, and bonus plans (short/long term incentive plans) so that they cannot be negatively impacted for at least one year from the closing date," it read, according to an employee lawsuit. "In the event of a layoff, any employee whose job is impacted would be eligible for severance."

Elizabeth Bruenig @ebruenig—Apr 14, 2022
if elon musk buys twitter and destroys it, history will have to acknowledge him as a morally complicated figure, capable of great evil but also great good

lesbian mothman @verysmallriver—Apr 14, 2022
i was on tumblr when yahoo had to sell it for $3 million after buying it for $1.1 billion. we were all a little bit responsible for losing yahoo a billion dollars. and with that type of collective effort, i believe we here on twitter can lose elon musk even more

Dewayne Perkins @DewaynePerkins—Apr 21, 2022
I started playing a game with myself where every time something annoys me on Twitter I immediately get off and force myself to do a Duolingo Spanish lesson, and bitch lemme tell you voy a ser bilingüe en poco tiempo porque todos ustedes son muy molestos

Dividend Hero @HeroDividend—Apr 25, 2022
Elon Musk was able to buy Twitter because he doesn't spend $5 on coffee everyday

Ned Miles @nedmiles—Apr 25, 2022
Can someone just tell me if I'm rich or fired please

Brooks Otterlake @i_zzzzzz—Apr 25, 2022
I hope a weird guy doesn't buy Email

Brittany Van Horne @_brittanyv—Apr 26, 2022
If we want to raise the money to buy twitter back and put it in the hands of the people, we're going to have to put on the greatest talent show this town's ever seen

CHAPTER 11

"Let's See What This Guy Can Do"

JP Doherty was on a surf trip in Southern California when he learned Elon Musk was buying the company. The forty-three-year-old engineering manager had a salt-and-pepper beard and a mischievous gleam in his eye. He talked, drove, and swore like a kid from New Jersey. On the rare occasions he got mad, it was at Bay Area drivers.

Every year, Doherty and his wife took their two kids, Rhys and Veronica, to a summer camp for kids with special needs. Both kids had autism. Rhys had mobility issues and sometimes used a wheelchair. The annual holiday was a highlight for the whole family, despite the hours of driving that it required.

Before joining Twitter, Doherty had spent twelve years as a professional musician, playing guitar and bass for artists like Blondie singer Debbie Harry. He'd toured all over the United States and Canada, playing to sold-out amphitheaters.

On the road, he was the fix-it guy, the one who stepped in to repair the amp that was on the fritz just before the show was set to start. His father had worked at AT&T and had taught him how to tinker with machines. He liked being a touring musician, but after he got married, he was ready for a change. So he'd quit that life and gone into tech. He started working at Twitter in 2012, left briefly to go to Apple in 2016, then returned four months later when he realized he missed his old job. By the time Musk came on the scene, he'd been at Twitter for the better part of a decade.

His role at Twitter wasn't too different from his days on the road. He was an engineering manager on Twitter Command Center (TCC), a twenty-four-person team that was the nervous system of the engineering organization, responsible for finding and fixing problems across the platform. If a bug impacted Twitter's users, it was TCC's job to find out what was causing it and stop it. If scrapers slowed Twitter's production systems, TCC would investigate how the scrapers got in, and block their access point. Doherty was still a fix-it guy, albeit one who made sure that a global communications platform stayed stable. He sat in front of huge monitors from morning to night, tracking Twitter's traffic as it pinged between data centers around the country. The only remnant of his days as a rock star was a signed Debbie Harry poster that hung on the wall of his home office.

Unlike some of his colleagues, Doherty wasn't reflexively anti-Musk. The guy seemed like kind of an asshole, but Musk also had something that Dorsey had lacked: the drive to make Twitter profitable.

Doherty was the sole breadwinner in his family. After Twitter went public, he'd been able to buy a small house in Alameda. Still, money was tight, and it irked him that Dorsey seemed morally opposed to doing anything that raised the company's share price.

Early in the pandemic, Doherty had driven across the country to pack up his mom's house and move her to California. She had Alzheimer's and required constant care. Between his mom's care and his kids' medical bills, Doherty's financial pressures were mounting.

Sure, Doherty shrugged when the news about Musk broke, *let's see what this guy can do.* He thought it was unlikely that he'd ever have to deal with Musk directly.

CHAPTER 12

"Temporarily on Hold"

On May 13, 2022, just eighteen days after Twitter accepted Musk's proposal, Musk announced without warning that the deal was "temporarily on hold." He told his followers that Twitter needed to prove that fake accounts were less than 5 percent of its users. Otherwise, the sale was off.

The news startled Twitter's board of directors. The deal was signed, and it did not allow Musk to temporarily put it on hold. What the hell was he playing at?

Throughout the negotiations, Musk had repeatedly emphasized that his desire to "kill the bots" was a primary reason he was buying Twitter. Now he was arguing that there were too many bots for him to buy the company. For years, Twitter had estimated that less than 5 percent of its monetizable daily active users (i.e., the number of accounts the platform could sell ads against) were bots or fake accounts. The company included this figure in its 10-K filings with the SEC, along with a disclaimer that read: "in making this determination, we applied significant judgment, so our estimation of false or spam accounts may not accurately represent the actual number of such accounts, and the actual number of false or spam accounts could be higher than we have estimated."

But Musk likely already knew this. The bots were a pretext—one that might help him wriggle out of the deal.

The markets had been sliding since mid-April and Twitter's stock was trading at $40.72 a share, down 9.7 percent from the day Musk made his initial offer. He claimed he'd been forced to sell 9.8 million Tesla shares, some for as low as $822.68, far below the $1,005 they'd been worth before he signed, to help fund the merger.

Musk liked to tell his staff that in every situation there were winners and losers and he never intended to be the loser. Unfortunately, in the case of Twitter, it looked like he might not have a choice.

As the fight escalated publicly, with Musk tweeting accusations that Twitter had lied about the number of bots on the platform, Parag Agrawal tried to do damage control. He earnestly explained the ins and outs of the company's spam estimate to Musk (and the rest of the world on Twitter). It was based on "multiple human reviews (in replicate) for thousands of accounts, that are sampled at random, consistently over time," from the pool of accounts that Twitter considered "monetizable daily active users."

Musk responded with a poop emoji.

CHAPTER 13

"Why We're Here"

The morning of June 16, Parag Agrawal kicked off the all-hands meeting that he'd been trying to schedule since April. Musk was running ten minutes late. Agrawal stalled for time as tweeps crowded into the Commons, an enormous cafeteria on the ninth floor of the San Francisco office. Finally, Musk logged onto the meeting from his phone. He was sitting in his living room in Austin, Texas. His white button-up was slightly undone, his hair more than slightly askew. A man in a black shirt (a friend? butler?) could be seen walking around in the background.

Agrawal thanked Musk for joining. Employees had submitted questions in advance and were eager to get started. The status of the deal was off-limits. Agrawal needed to assume that the acquisition was moving forward regardless of Musk's recent antics. But everything else—Musk's political views, remote work, compensation—was fair game.

Leslie Berland, Twitter's chief marketing officer, eased Musk into the Q&A with a softball. "Why do you love Twitter?" she asked.

"Well, let's see. I find, like, I learn a lot from what I read on Twitter, and what I see in the pictures, videos, text, and memes that people create," Musk said. "I also find it's a great way to get a message out over the phone, when I want to say something and make an announcement, I think Twitter's the best way to do that."

He continued to ramble, joking that "some people use their hair to express themselves, I use Twitter."

Employees were growing concerned. "I actually turned to a stranger in the Commons and said, 'What is happening right now?'" one employee recalls. Wasn't this guy supposed to be a genius?

Randall Lin thought people were overreacting. "With Elon you kind of know what you're going to get," he says.

Berland asked Musk how Twitter employees could earn his trust and how he planned to earn theirs in return. Musk demurred. "It's, like, if somebody is getting useful things done, then that's great," he said vaguely. "But if they're not getting useful things done, then I'm, like, 'OK, why are they at the company?'" Agrawal might have hoped this meeting would comfort Twitter workers, but Musk seemed to have no interest in doing that.

Berland's last question was whether Musk planned to take the title of CEO. Musk responded that he didn't get hung up on titles. His role at Tesla was "techno king" as well as CEO, and his chief financial officer was called the "master of coin" in addition to being the CFO.

Then he laid out what he called his unifying philosophy. As far as employees could tell, it had nothing to do with Twitter.

"We should take the set of actions most likely to extend the scope, scale, and life span of consciousness as we know it," he said. "What sort of actions improve things at a civilizational level and improve the probable life span of civilization? Civilization will come to an end at some point, but let's try to make it last as long as possible. And it would be great to understand more about the nature of the universe. Why we're here, meaning of life, where are things going, where we come from? Can we travel to other star systems and see if there are alien civilizations? There might be a whole bunch of long-dead, one-planet civilizations out there that existed five hundred million years ago. Think about the span of human civilization from the advent of the first writing, it's only about five thousand years."

Many employees were stunned. "This is someone who is either quite stupid or hasn't given this much thought," a former engineer said. "But either way he is not taking this seriously, and he holds the company and the product and us, the employees, in contempt."

Another engineer manager left the meeting and told his team that if they ever ran a meeting like that, they'd be fired.

"He did not land the plane," a former executive noted dryly.

CHAPTER 14

"Hide-and-Seek"

Parag Agrawal arrived in Sun Valley, Idaho, on July 4, 2022, on a private jet. The picturesque mountain town (population: 1,814) was undergoing its annual transformation into a summer camp for billionaires, as the exclusive conference hosted by boutique investment bank Allen & Co. kicked off at the ritzy Sun Valley Lodge.

It was a dramatic summer for masters of the universe. The financial markets were still backsliding. Rumors circulated that Bob Chapek, the embattled CEO of Disney, might run into Bob Iger, his predecessor, former mentor, and current enemy. Yet, as attendees hobnobbed with Bill Gates and Warren Buffett, the conversations kept straying back to one thing: Elon Musk's contested acquisition of Twitter.

Musk was headlining the event, but he was nowhere to be seen for the first two days. Occasionally, reporters spotted Agrawal, Bret Taylor, and Twitter's CFO, Ned Segal, strolling around the grounds, looking suspiciously relaxed, but none of them would answer a single question about the deal.

Finally, on the evening of July 7, Musk's jet touched down in Idaho. The following morning, people's phones started to buzz. Quinn Emanuel partner and longtime Musk attorney Alex Spiro, who has also represented the actor Alec Baldwin, the hip-hop icons Jay-Z and Megan Thee Stallion, and former NFL tight end and convicted murderer Aaron Hernandez, had delivered (along with Musk's other attorneys) an eight-page letter to

Twitter notifying the company that Musk was terminating the merger agreement. He claimed the company was misrepresenting the number of users who were bots. “First, although Twitter has consistently represented in securities filings that ‘fewer than 5%’ of its [monetizable daily active users] are false or spam accounts, based on the information provided by Twitter to date, it appears that Twitter is dramatically understating the proportion of spam and false accounts represented in its [monetizable daily active user] count,” the letter said.

The following day, Musk sat onstage across from Sam Altman, CEO of the artificial intelligence firm OpenAI, which Musk had backed early on. The auditorium was packed. Everyone, from Altman to the audience members, wanted to know what was going on with the Twitter deal, but Musk refused to answer. He talked about the platform’s shortcomings—the prevalence of bots, the mistakes it had made in banning former President Trump—at one point even turning to the audience to ask whether they believed 95 percent of Twitter’s users were authentic, as the company claimed. Heads turned to stare at Agrawal, Taylor, and Segal, but the executives managed to look neutral. Musk was putting on a show, but they knew Twitter held all the cards. Musk had signed away his right to due diligence with an airtight purchase agreement.

Three days later, on July 12, 2022, Twitter made its move, suing Musk to force him to buy the company.

“Having mounted a public spectacle to put Twitter in play, and having proposed and then signed a seller-friendly merger agreement, Musk apparently believes that he—unlike every other party subject to Delaware contract law—is free to change his mind, trash the company, disrupt its operations, destroy stockholder value, and walk away,” the company’s lawyers wrote in the complaint. The suit alleged that Musk had signed a binding agreement to buy the company (which he had) but was now trying to get out of that agreement because market conditions were grim (which they were). “Rather than bear the cost of the market downturn, as the merger agreement requires, Musk wants to shift it to Twitter's stockholders," the complaint read.

Twitter's lawyers pointed out the obvious discrepancy in Musk's argument. Namely, that the reason he'd wanted to buy Twitter in the first place was to, in his own words, "kill the bots," and now he was using the bots as a pretext to kill the deal.

"Musk wanted an escape. But the merger agreement left him little room," the lawyers wrote dramatically. "With no financing contingency or diligence condition, the agreement gave Musk no out absent a Company Material Adverse Effect or a material covenant breach by Twitter. Musk had to try to conjure one of those."

In fact, Musk had tried to conjure both.

First, he'd accused Twitter of lying about the number of bots on the platform. This argument didn't look very viable. The company had always been careful to specify that the 5 percent figure was an estimate. Even if Musk could prove the company was lying, he'd need to show that the lie had a material adverse effect on Twitter's business.

Musk had another option. As part of the merger agreement, Twitter had agreed to give Musk data and information "for any reasonable business purpose related to the consummation" of the deal. Musk needed to show Twitter was withholding information.

Already, the two sides had met multiple times to discuss how exactly Twitter calculated the percentage of its monetizable daily active user accounts that were bots or spam. Twitter had handed over calculations, data sets, and proprietary information. But Musk and his lawyers never seemed satisfied. If anything, their attitudes grew chillier the more Twitter tried to play ball. On May 13, the day Musk announced the deal was on hold, Twitter's CFO, Ned Segal, called into a Zoom meeting while he had Covid. As Musk's bankers grilled the Twitter executives, Segal turned away from the camera to cough. "We get it, Ned, you're sick," one of the bankers hissed, her voice dripping with vitriol.

Throughout the spring and early summer, an absurd back-and-forth played out, where Musk's team asked for enormous amounts of data, and Twitter complied, and then Musk's team asked for more.

Finally, after weeks of playing cat and mouse, the two parties agreed to meet in the Chancery Court in Delaware to hammer out their arguments in front of a judge. Musk's lawyers wanted to delay the trial, but the judge, Kathaleen St. J. McCormick, sided with Twitter, and scheduled a speedy five-day trial for October. "The longer the merger transaction remains in limbo, the larger the cloud of uncertainty casts over a company and greater the risk of irreparable harm to sellers and to the target itself," McCormick wrote in her brief.

Musk's lawyers continued asking Twitter for information. But McCormick ruled that some of the requests were absurdly broad. She seemed to see Musk's strategy for what it was: a stalling tactic.

Undeterred, Musk filed a countersuit in July, reiterating his claims that the company had lied about the number of bots on the platform. "Twitter played a months-long game of hide-and-seek to attempt to run out the clock before the Musk Parties could discern the truth about these representations, which they needed to close," the suit alleged.

In the lawsuit, Musk's lawyers also claimed that their client had become "increasingly concerned in recent years with the company's direction and poor user experience, given the flood of misinformation, scams, and other undesirable content he regularly sees."

Already, the back-and-forth was hurting Twitter's share price. In August, the stock was trading at $38.75 a share, down from $49.02 at the end of April. Segal emailed employees to warn them that their annual bonuses would likely be half of what they'd hoped at the start of the year. As the fight dragged on, employees weren't sure which outcome looked worse: Musk buying the company, or Musk walking away and leaving the stock price in the dirt.

CHAPTER 15

"Some Things Are Priceless"

Parag Agrawal hadn't expected his chief security officer to become such a problem. He'd fired Peiter "Mudge" Zatko in a wave of executive layoffs in January 2022, soon after he became CEO, but it wasn't until the height of summer, with the lawsuit against Elon Musk reaching a boiling point, that the decision came back to haunt him.

Jack Dorsey had personally recruited the veteran hacker to come work at Twitter after a humiliating cyberattack in 2020 in which teenage scammers called Twitter employees and tricked them into handing over their credentials. The kids had taken over accounts belonging to figures including Joe Biden, Barack Obama, Kanye West, Elon Musk, and Jeff Bezos to promote a fake bitcoin scheme.

In the wake of the attack, Twitter employees were thrilled that the leader of the famed hacking collective Cult of the Dead Cow was going to be joining their ranks. "I was psyched," said Ian Brown, a senior engineering manager. "He's got a great brand in InfoSec."

But it didn't take long for Brown and his colleagues to become suspicious of their new security chief. Shortly after Mudge took over, Brown received an email from a Twitter VP asking him to send a detailed report outlining all the versions of Twitter's operating systems and kernels (a component of an operating system that connects hardware and software) currently in use to a personal email address belonging to a man in Texas. Brown worried it might be a phishing scam. Every version of the operating system had specific flaws. Telling someone outside the company—via an unencrypted email—exactly which versions Twitter was using was like handing them a map to the company's security vulnerabilities.

Brown got on a video call with a senior engineering leader and asked what was going on. The leader responded that Brown needed to do what Mudge said. Brown reluctantly sent the report.

His opinion of Mudge deteriorated further when the security chief deprioritized two major security initiatives to encrypt Twitter's servers (a project called "encryption at rest") and keep its operating systems up-to-date—even going as far as to take them off the UPL. "I don't think it was necessarily malice, but it was certainly a lack of effort to understand Twitter's infrastructure," Brown said.

Another security engineer who worked under Mudge recalls the executive twice falling asleep during executive review meetings. The behavior did not instill confidence. Few in the security organization were sad to see him go in January.

After Mudge was fired, settlement talks dragged on for months. In June, Twitter agreed to pay him $7 million. Two months later, Mudge filed an eighty-four-page whistleblower complaint to the Federal Trade Commission in which he accused his former employer of having "extreme, egregious deficiencies" in its security apparatus. He heaped a hefty amount of blame on Parag Agrawal.

Mudge claimed that over 50 percent of the servers in Twitter's data centers were operating with noncompliant kernels or operating systems—an allegation that shocked Ian Brown. Mudge seemed to be conflating kernels—the core of an operating system—and the operating system (OS) itself in a single metric, which was "insane and very inaccurate," Brown says. Plus, Twitter automatically updated its operating systems every two weeks. The level of noncompliance Mudge was asserting, at least for the OS, didn't make sense.

Mudge also claimed that many of Twitter's servers were "unable to support encryption at rest"—a project he'd personally deprioritized, along with one other executive, according to Brown. "It was really frustrating to see that show up in the report when Mudge did more to prevent us from working on this than anyone else, which is saying a lot," Brown tells me.

Mudge's former team was outraged about the allegations. But another aspect of the whistleblower complaint grabbed most of the attention.

On May 16, after Musk announced the deal was on hold, Agrawal had responded publicly and said that Twitter executives were "strongly incentivized to detect and remove as much spam as we possibly can, every single day." Mudge called bullshit. "Agrawal's tweet was a lie," he said in the whistleblower complaint. "In fact, Agrawal knows very well that Twitter executives are not incentivized to accurately 'detect' or report total spam bots on the platform."

He claimed that "senior executives earn bonuses not for cutting spam, but for growing mDAU" (monetizable daily active users). This was a fancy way of saying that Twitter's leadership was motivated to grow the number of users that Twitter could show ads to—a goal shared by nearly every executive at every big tech company in Silicon Valley.

Contrary to Musk's argument that Twitter included fake accounts in its number of monetizable daily active users, Mudge said that Twitter was doing "a decent job excluding spam bots and other worthless accounts from its calculation of mDAU."

But he still believed executives could be doing a much better job eliminating bots on the rest of the platform.

Musk's lawyers subpoenaed the former security chief, eager to bolster their case before the trial started in October. From the outside, the whistleblower complaint looked like the break Musk needed to torpedo his acquisition of Twitter. If bots were an excuse to get out of the deal, here was surprising evidence that could potentially delay the sale.

Mudge's complaint set off alarm bells at the Federal Trade Commission, which was experiencing a seismic shift under the leadership of chair Lina Khan, who was nominated by Joe Biden in 2021. For the last hundred years, the agency, which was tasked with protecting consumers and promoting competition, had taken a more moderate approach to regulating big business, with a major focus on price. Khan believed that this approach was outdated, particularly in regards to big tech. As a student at Yale Law School, she'd written a paper titled "Amazon's Antitrust Paradox" that argued that focusing on price failed to account for the harms of big tech monopolies. "We cannot recognize the potential harms to competition posed by Amazon's dominance if we measure competition primarily through price and output," she wrote.

Khan indicated that she was willing to take big swings even if it meant losing in court. "Even in cases where you're not going to have a slam-dunk theory or a slam-dunk case, or there's risk involved, what do you do?" she asked in an interview with *The New Yorker*. "Do you turn away? Or do you think that these are moments when we need to stand strong and move forward? I think for those types of questions we're certainly at a moment where we take the latter path."

Since 2011, Twitter had been operating under a consent order with the FTC over allegations that it failed to adequately protect user privacy. In May 2022, the Justice Department, on behalf of the FTC, filed a complaint accusing Twitter of violating the consent order, after it took users' phone numbers and email addresses, supplied for account security, and let advertisers use that information to target ads. The company paid a $150 million civil penalty and agreed to a strict set of provisions to ensure it wouldn't happen again.

"When you enter into a consent order with the FTC, the penalty itself is by far the lowest part of that cost. The conduct provisions, the requirements regarding what you have to do moving forward, can be much more costly, if you do them right," says Kim Phan, a partner at Troutman Pepper who clerked at the FTC and now counsels companies on federal and state privacy and data security regulations.

At Twitter, these conduct provisions boiled down to a series of checks and balances called the "flyway" process. Before a new feature rolled out, employees had to meticulously document how it would impact user privacy. First, project managers wrote up a proposal, discussing the privacy and security implications of the initiative. Then they put together a technical design document and opened a security ticket to make sure no personally identifiable information would be misused. The process involved hundreds of employees across multiple teams, including privacy and engineering. In the whistleblower complaint, Mudge accused Twitter of failing to follow the flyway process in a number of key ways.

For example, when a Twitter user submitted a request asking that their data be deleted, the company was supposed to find and delete their information off its servers, quickly and ideally automatically. Mudge said Twitter had deliberately misled the FTC on this account. "[W]hen the FTC asked Twitter whether it fully deleted the data of users who left the service, Twitter deliberately misled the FTC by stating those accounts were 'deactivated,' even when the data was not fully deleted," he wrote.

Twitter was actively working to fully automate the process of finding and deleting user data, but some parts were still being handled manually, meaning they took longer than they should have, and some data lingered on Twitter's servers even after a user submitted a deletion request. Internally, the automation initiative was known as "Project Eraser." It was on track to be completed in the fall of 2022.

After Mudge's whistleblower complaint became public, the FTC opened a probe into Twitter's privacy and security practices. The commission was well aware of Musk's pending acquisition. Regardless of who owned the company, regulators were on high alert.

Employees who'd been wary of Musk's initial offer were even more on edge now. He had spent the summer bad-mouthing Twitter and its executives. And his personal conduct was getting more lurid. In July, *Insider* published a scoop that Musk had fathered twins with Shivon Zilis, one of his top executives at Neuralink, via in vitro fertilization. (Grimes would only discover the existence of these kids from the media.)

It seemed clear that Musk had little regard for what was considered appropriate behavior in the workplace. But having children with a subordinate? Musk explained the decision as part of his mission to save the human race.

"Doing my best to help the underpopulation crisis," Must tweeted after the story dropped. "A collapsing birth rate is the biggest danger civilization faces by far."

Some Twitter workers were confused. "People were honestly befuddled," one told me.

"At that point, I was just stupidly hopeful the deal wouldn't happen," said another.

At higher levels of the company, Twitter leaders were absolutely disgusted. "He got one of his female leaders pregnant. Are you kidding me?" a female executive told me.

But she was merely feigning shock. At this point, nothing about Musk's behavior could surprise anyone.

All throughout the summer, Musk's advisers had been trying to convince him that he didn't have a strong case to get out of the deal. Mudge's allegations had been a gift out of the blue, but even then, the lawyers had to see that the most they could get out of Mudge's complaint was a delay. Finally, on October 4, 2022, with the court date looming and

Musk's text messages going public in discovery, the CEO accepted his fate. Just as quickly as he'd decided to buy Twitter—and then decided he no longer wanted to buy Twitter—he decided he *had no choice* but to buy Twitter. He agreed to acquire the company for the original terms of the agreement, as long as Twitter dropped the lawsuit.

Musk would later say he felt like he'd bought a warehouse of goods thinking only 5 percent of them were broken, only to look inside the warehouse and see that the number was closer to 25 percent. Still, publicly losing at trial and being forced to buy the company was unacceptable. If Musk had to buy Twitter, it would be on his terms. The company wasn't worth $44 billion, the price he'd declared himself, but there wasn't much he could do about that now.

"Some things are priceless," he later reasoned.

Part II

HELLSITE

CHAPTER 16

"Let That Sink In!"

On Wednesday, October 26, 2022, Elon Musk strolled into Twitter's headquarters carrying the porcelain top of a sink. The deal had to close in two days, otherwise the case would go to trial. Musk wanted to come in early to look around. He used the opportunity to shoehorn in a pun: "Entering Twitter HQ—let that sink in!"—in a moment made to go viral, rather than to comfort the employees who wondered what it would be like to work under their volatile new leader. Twitter had seventy-five hundred employees. The sink tweet got 1.4 million Likes.

Musk arrived with an entourage. Dozens of engineers from Tesla, SpaceX, Neuralink, and The Boring Company were seen around the office in the days to come, along with a host of advisers: Jason Calacanis and Antonio Gracias; the former chief operating officer of PayPal David Sacks; Musk's lawyer Alex Spiro; a venture capitalist named Pablo Mendoza; and the wealth manager Jared Birchall, among others. Officially, the group was called the "transition team." But Twitter employees just called them the Goons.

Randall Lin was working at The Perch, an upscale coffee bar on the tenth floor of the main office, when the room started filling up with people. The engineer craned his neck to see what was going on. Twitter was never this busy. Then he spied the unmistakable form of Musk wearing a black T-shirt and his signature black cowboy boots, a silver chain dangling from his neck. He was standing with Leslie Berland, Twitter's polished chief marketing officer. Musk's two-year-old son, X Æ A-Xii, toddled closer to Lin, snacking on Goldfish crackers. *Well,* Lin thought. *This day just got a lot more interesting.*

Berland was based in New York but had flown to San Francisco on a private jet that morning (a trip that cost Twitter $103,850, which Twitter later refused to pay) to help squire the incoming CEO around the office. "Elon is in the SF office this week meeting with folks, walking the halls, and continuing to dive into the important work you all do," she wrote in a company-wide email. "If you're in SF and see him around, say hi! For everyone else, this is just the beginning of many meetings and conversations with Elon, and you'll all hear directly from him on Friday."

The tone was purposefully cheerful. Six days earlier, *The Washington Post* published a story alleging Musk planned to cut nearly 75 percent of the company if the deal went through. Anxiety was running high, and yet Musk had undeniable star power—layoffs or no, people were excited to meet him.

At the coffee bar, a harried food services manager shooed the regular barista out of the way and started making Musk's drink: a coffee, microwaved please. Above his head, a glowing silver sign read #LOVEWHEREYOUPERCH—a play on the #LoveWhereYouWork hashtag that epitomized Twitter's office culture.

Musk leaned back against the bar, sipping his drink, as a crowd of employees gathered around. Some held out their phones to snap a picture. "Employees were basking in Musk's aura," a former engineer recalls.

A product manager named Esther Crawford sidled up to Musk to introduce herself. She had ideas on how to make Twitter a more attractive platform for creators and anticipated Musk would be a receptive audience. "I saw [Musk] as the guy who built incredible and enduring companies like Tesla and SpaceX, so perhaps his private ownership could shake things up and breathe new life into the company," she later said. She told

Musk she wanted to meet one-on-one to discuss her ideas. Musk, who at 126 million Twitter followers was one of the company's biggest creators, seemed receptive.

Crawford, an early YouTube vlogger who'd grown up in a Christian cult and became internet famous by posting videos about her weight-loss journey, joined Twitter in 2020 after the company acquired her startup. Years before, she'd live-tweeted her contractions while giving birth. She got married, then divorced. She married again at Burning Man in 2022. She wrote about treating her depression with psychedelics. *Vanity Fair* detailed her "sexual experiments and open relationships" in an article about Silicon Valley orgies. She was well-liked by her team, but her go-getter attitude had put her at odds with some members of Twitter's leadership.

Crawford was frustrated with how slow and bureaucratic Twitter had become. Once, she'd watched a colleague spend an entire month trying to get approval to reach out to a group of content creators. "He went through three layers of management and six different functional teams. In the end, four executives were involved in the approval," she later wrote. "It was insanity." (For transparency, I reached out to Crawford repeatedly for comment, but she never responded.)

More people filtered in as word of Musk's presence spread. Two of Lin's friends arrived at the office. They didn't work at Twitter, but they'd noticed the sink tweet and wanted to see Musk in person. Lin put them on the guest list, certain their day passes would be denied, but to his surprise they'd sailed right past security. Now all three watched as Musk held court.

One of Lin's friends tweeted a photo. How could he resist? The fact that he'd gotten into the office, that he was part of this moment the whole world seemed to be watching, was a flex.

"Patriots in control," he said. "[M]y vibe check is that he's doing the rounds to demonstrate legitimacy as a ruler to the smallfolk. Dude brought his baby to kiss and generate good energy. Succession ass moves happening."

Later that day, Lin got a call from security. "Who did you bring into the office?" they asked. "How did they get in? What are their goals?"

Shit, Lin thought. *That was stupid.*

Lin reassured the security team that his friends didn't mean any harm. They just wanted to be in Musk's orbit. But internally he was kicking himself. Technically, it was security's job to vet people on the guest list, not his. And yet, if the acquisition went through, Musk was poised to change everything Lin hated about Twitter's culture. Lin wouldn't—he *couldn't*—jeopardize being a part of that.

Back at the coffee bar, a female employee took the opportunity to ask the question that was on everyone's mind. "I know we're all superexcited to meet you," she said. "But I think the real question everyone's thinking is: Are you going to fire 75 percent of us?"

Musk looked up at the ceiling, puzzled. "You know, I'm not actually sure where that number came from, because . . . no," he said finally. "It didn't come from me."

Emmanuel Cornet, known to friends as Manu, was working at a WeWork in downtown San Francisco when he saw Musk's tweet about the sink. *I should be there*, he thought. He liked being around people, but found he could focus better when those people weren't his colleagues. Musk's arrival was a worthy exception.

The French engineer had worked at Google for fourteen years before taking a job at Twitter in 2021. He'd noted the differences early on—Twitter outsourced a lot of jobs to smaller tech companies, Google liked to build tools in-house; Twitter's engineering team had a lot of women, Google's had way more men; Twitter employees collaborated on Slack, Google employees on Gmail—and decided he preferred Twitter's culture. "What makes the company's culture is 100% us," he wrote confidently on his blog shortly after Elon announced his intention to buy Twitter. "If we're not going anywhere, neither is the culture."

Cornet, a polymath whose wry cartoons had been featured in *The New York Times*, was known as a bit of a troublemaker. In 2021, after the newspaper broke the story about Twitter's controversial design chief Dantley Davis, Cornet published a cartoon that seemed to make light of the allegations. In the first panel, Jack Dorsey hands Davis the iconic bird logo and asks him to shake up the company's culture. In the next panel, Davis vigorously shakes the bird upside down. "Erm, that's a little too much," cartoon Dorsey says.

Twitter's HR team asked Cornet to take the drawing off Slack as well as his public Twitter profile. "I took down the internal post, but I took issue with the fact that they also asked me to take down my public Twitter post (which should have been outside of their authority)," he wrote on his blog. Dorsey, whose Twitter bio read "#bitcoin and chill," apparently hadn't known about the request and sent him a direct message asking for more information. The conversation tapered off, but Cornet was never bothered about his cartoons again.

Over the summer, as Musk tried to back out of the deal, Cornet's optimism about the acquisition faded, but only slightly. He didn't like what Musk was saying about Twitter, but he also doubted that a single person could have *that* big of an effect on the company culture. At least some of Musk's antics had to be a performance. Cornet was willing to give him a chance.

Now, Cornet walked the block to the office, where on a whim he printed out a cartoon to give to his incoming boss. The engineer didn't want to present Musk an overly positive cartoon. His colleagues would think he was a bootlicker, especially since part of his brand was being playfully critical of Twitter's management. But he also didn't want to get fired by giving Musk a cartoon that pissed him off. He chose a middle ground: a simple drawing showing two men standing by a shelf filled with company logos. Twitter's bird logo lay smashed on the ground. "You break it, you buy it!" one man yelled. *Perfect*, Manu thought. *Playful but not blatantly insubordinate.*

Once Cornet spotted Musk, he got shy. The executive was surrounded by adoring fans. Cornet didn't want to push past everyone to give the CEO a silly cartoon. He tapped Berland on the shoulder to ask if she'd do it for him. "No, you give it to him," she said encouragingly.

A few minutes later, Berland was showing Musk around the office and happened to walk by Cornet's desk. The engineer took a deep breath and handed Musk the cartoon.

"For Elon Musk, I hope you don't mind a 'court jester' at Twitter, or you'll have to get me fired," he wrote in the corner. He signed it with a mischievous-looking smiley face. Dorsey and Agrawal would have appreciated the drawing. Cornet hoped Musk would feel the same.

But Musk simply looked at the paper, as if confused about why Manu had handed it to him. "Well . . . I bought it anyway," he said, stone-faced. Then he walked away, leaving the engineer feeling spectacularly deflated.

I need to be in the room, Lin thought as he walked across the glass sky bridge. He'd heard the rumors. Musk was meeting with small groups of employees to understand Twitter's business. Lin wanted to be one of those employees.

Twitter's San Francisco headquarters was split between two buildings. The main office, on Market Street, was a stately art deco building from the 1920s, with Mayan-inspired panels on the roof. On the corner, a massive TWITTER sign hung vertically, announcing the company's presence to passersby below.

Like most of his colleagues, Lin worked in the main office on Market Street. But, as if to signal that they weren't all one happy family, Musk and his team set up camp in Twitter's second office, around the corner on Tenth

Street. This space was a modern homage to concrete and metal, with floor-to-ceiling windows and cavernous conference rooms for formal meetings. It fit Musk's vibe better than the main office, with its yoga rooms and cheerful Twitter-branded decor.

Lin was used to moving around Twitter's offices freely. The buildings didn't have strong security, as he'd found out when his friends had been let in. But when he arrived at the office on Tenth Street, Musk's security detail stopped him. "Do you have a meeting?" they asked. Lin admitted he did not. The security guards turned him away.

The following day, Lin was eating lunch in the Twitter cafeteria when he spotted one of the Tesla engineers. The two got to talking. "Why do half the engineers have no commits for the last 30 days?" the man asked, referring to the process of adding code to the code base. Something clicked in Lin's mind. *They will be making tough decisions with very limited information*, he realized. *Which means I need to be providing that information.*

CHAPTER 17

"I Understand How Computers Work"

The afternoon of October 26, Yao Yue was sitting at her desk when she got a message telling her that Elon Musk wanted to meet. He had forty-eight hours to close the $44 billion deal. It was time to figure out what he was buying.

The meeting is supposed to be a secret, she was told. Yue rolled her eyes. It was hard to take all this cloak-and-dagger stuff seriously. Twitter employees talked. Transparency was a core part of the company culture. If Musk thought tweeps were going to show him more loyalty than they showed one another, he clearly had no idea who he was dealing with.

Still, Yue was curious to see what Musk was like in person. He called himself an engineer and bragged about his technical genius. Yue wanted to put it to the test. As much as this meeting was about Musk understanding Twitter's tech stack, it was also about Yue understanding Musk.

Since April, Yue had made the conscious decision to separate herself emotionally from her job. She loved Twitter and cared deeply about her colleagues, but she was also a pragmatist. If Twitter's board of directors wanted to sell the company to Musk, she couldn't stop them. There was no point in getting all spun up about a decision she had zero control over.

Yue had grown a lot since she'd started working at Twitter in 2010.

Early on in her career, she'd spent a lot of energy convincing other people she was right. She wasn't combative, but if someone disagreed with her, she couldn't just let it go.

One night, months after she'd started, Yue was working late at the office when she got into an argument with one of her closest colleagues, an engineer named Manju. Manju was only a year older than Yue, but he had far more industry experience. Yue liked to push him—she wanted to prove that even though Manju had spent more time in tech, she was just as good an engineer.

That evening, however, it seemed like Manju had had enough. He asked Yue to step into a conference room. "You need to stop trying to prove you're right," he told her. "You're smart and you have good ideas, but you're making it hard for people to collaborate with you."

Yue wanted to argue. The words were on the tip of her tongue. *I'm not trying to prove I'm right! I love collaboration!* But she bit her lip as the truth of what Manju was saying sunk in. The truth was, despite how much she'd accomplished, she still felt insecure, and that insecurity made her argumentative. She had to learn how to shut up and admit when she was wrong.

After that, Yue made a concerted effort to listen and cop to her mistakes. She was grateful that Manju cared enough to give her honest feedback. His intervention made her a better programmer. Over the years, Yue's career at Twitter flourished as she built core aspects of Twitter's infrastructure and rose up the ranks to become a principal engineer. If an employee wanted their code to run fast, Yue was the person to talk to.

Which brought her to that afternoon in October, and a conference room in San Francisco that had once been used by Parag Agrawal but now housed a handful of high-ranking Twitter employees, all waiting apprehensively for Musk.

A few minutes passed before Musk hurried in and took a seat. After a short round of introductions, the tweeps started going through their slides, which they'd adapted from a new-hire training called "How Twitter Works." The presentation was fairly easy for software engineers to understand but hopelessly complex to everyone else.

Under the hood, Twitter was a series of "microservices" that each performed a specific function. For example, there was Tweetypie, which handled tweets; Gizmoduck, which managed user accounts; and Social Graph Service (SGS), which was responsible for information regarding which accounts a user followed or unfollowed.

When a Twitter user posted a tweet, that tweet was sent to Tweetypie for storage. If that user changed the photo on their profile, that change would be recorded in Gizmoduck. And then, when it came time to relay their tweets and profile to other users on Twitter, a fanout service was responsible for ferrying the tweet over to the home timeline and search.

Each service was connected to a database and caching (a term for high-speed storage) layer. Yue had built Twitter's caching system, which was called "distributed caching." It helped Twitter serve a massive audience with as few servers as possible.

A downside of the microservice architecture was that it was complex. But the alternative—a monolithic service—was harder to scale.

David Sacks, who had worked with Musk at PayPal, slipped quietly into the room. Sacks was a polarizing figure in Silicon Valley. In 2012, shortly before he sold his startup Yammer to Microsoft for $1.2 billion, Sacks threw himself a lavish fortieth birthday party. Guests were told not to share photos or videos. But Snoop Dogg, who'd been hired to play at the event, broke the rules, and posted a photo of Sacks wearing a cravat and waistcoat, in the style of Marie Antoinette. The party slogan was, unironically, "Let Him Eat Cake."

As the engineers discussed Twitter's architecture, Musk waved Sacks out of the room. "David," Musk said, "this meeting is too technical for you." (Sacks, through a PR firm, disputed this story.)

Yue filed that away for later.

Then, as Yue painstakingly explained Twitter's infrastructure, Musk interrupted to ask about cost. The current setup seemed pricey. How could Twitter save money?

Yue tried to explain that modern computer architecture had quirks. You couldn't cut costs without thinking through the design of the whole system. "I was writing C programs in the '90s," Musk said dismissively. "I understand how computers work."

Yue filed that away for later, too.

Now Musk was ready to move on. "We really should be able to do long-form video and attract the best content creators by giving them a better cut than YouTube," he said.

The suggestion caught employees off guard. Long-form video was *technically* possible. But these were infrastructure people—might Musk want to speak with the media team instead? He didn't seem interested in this suggestion.

After the meeting, Yue went to Jay Sullivan, Twitter's head of product, to tell him how it had gone. "Elon seems like a product manager, not an engineer," she said. It wasn't necessarily a bad thing—lots of CEOs were good product managers and left technical decisions to their developers.

But Sullivan warned her to stay quiet. "Don't tell Elon that," he said. "He's going to be really mad if you tell him that."

Later that day, Yue recorded a voice memo so she wouldn't forget the interaction. "The session felt exploratory," she says on the recording, sounding slightly amused. "I don't think Elon was quite technical enough . . . he wasn't particularly interested in the technical details of how certain features work. He was asking reasonable questions and clearly had very deep concerns about the cost of running the business and was eager to explore avenues to reduce costs and make ends meet."

A lot of things would surprise Yue in the coming weeks. But Musk's focus on cost cutting wasn't one of them.

CHAPTER 18

"Content Moderation Is a Product"

With the deal poised to finally close, advertisers were growing increasingly concerned that a free speech absolutist was going to take one of their favorite marketing channels and turn it into an unmanageable hellscape.

Historically, Twitter had sold itself as a prestige platform for marketers. It didn't have the targeting capabilities of Meta or the reach of Google. But it had an elite audience made up of journalists, celebrities, and tech influencers. If a brand wanted to create buzz around a product, Twitter was the place to be.

This value was predicated on Twitter being a safe place to advertise. The company had an entire brand safety team dedicated to ensuring a good advertising experience—one where ads didn't run next to adult content, violent images, or hate speech. It wasn't a perfect system, but it worked.

Or, at least, it *had* worked. No one knew exactly what Musk meant when he talked about free speech. But anyone who understood the dynamics of social media knew that loosening the reins of what could be said would also make Twitter less predictable. For those who saw Twitter as a "global town square" where all ideas could be discussed and debated, Musk's vision was a dream; but advertisers—the people who constituted Twitter's core business—could not be less interested in a free-for-all. All any advertiser wanted was consistency and a basic transaction: to pay money and hope that an ad pushing a product, service, or brand might show up next to a benign tweet from LeBron James. Even the off chance that an ad might appear near anything resembling hate speech was an advertiser's nightmare.

"Advertisers play an underappreciated role in content moderation," says Evelyn Douek, a professor and speech regulation expert. "So much of the content moderation discourse has always been a highfalutin discussion on free speech, on safety versus voice. But content moderation is a product, and brand safety has always been a key driver in terms of how these platforms create value."

All this sounded a lot like censorship to Musk. Once he owned Twitter, he planned to roll out a subscription product. Free Twitter from ads, and he'd free it from the constraints of advertisers.

Before this could happen, however, Musk needed the ad industry on his side. As soon as the deal closed, Twitter would be laden with debt. That, and the surplus of employees, meant Twitter was about to hemorrhage money. If advertisers fled, the company would be in dire financial straits.

On Thursday, October 27, Musk posted a conciliatory letter to try to coax advertisers to stick around.

"There has been much speculation about why I bought Twitter and what I think about advertising," he wrote. "Most of it has been wrong. The reason I acquired Twitter is because it is important to the future of civilization to have a common digital town square, where a wide range of beliefs can be debated in a healthy manner, without resorting to violence."

Musk blamed traditional media for fueling polarization in the "relentless pursuit of clicks." He said he hadn't bought Twitter to make money but because he loved humanity and wanted to help it.

Finally, he addressed the elephant in the room. "Twitter obviously cannot become a free-for-all hellscape, where anything can be said with no consequences!" he wrote. ". . . Fundamentally, Twitter aspires to be the most respected advertising platform in the world that strengthens your brand and grows your enterprise."

Not everyone was convinced. "Musk can make empty gestures in the direction of civility all he wants, but the fact is that he's been transparent about his intention to re-platform Twitter's most toxic users and to gut content moderation," said Andrew Graham, founder of the brand consultancy Bread & Law. "You don't 'debate' whether, for example, vaccines work, or whether people of certain races, ethnicities, or gender identities deserve to live. And you don't improve the health of a conversation by inviting bigots, liars, and grifters into it."

The next day, General Motors, a Tesla competitor that typically spent $1.7 million a month on Twitter ads, paused all its marketing campaigns on the platform, apparently unconvinced by Musk's promises. General Mills, Audi, and Pfizer slowed down their advertising spend indefinitely. Many more brands drastically reduced their advertising spend, effectively torpedoing Twitter's revenue. Just as Musk had feared, Twitter was already underwater.

CHAPTER 19

"The Bird Is Freed"

On Thursday, October 27, 2022, Yao Yue biked with her two young children to the office for a Halloween party called "Trick or Tweet." The ninth floor of the office was decked out with miniature pumpkins and fake spiderwebs. Outside on the balcony, a clown blew bubbles. Another performer walked around dressed like a scarecrow. He looked like he was being tailed by a handler. Employees started whispering to one another. *It couldn't be . . . was it? Was this Musk dressed up in a costume?* It turned out to just be a performer.

Yue tried to keep her spirits up. While the protracted back-and-forth over the summer had felt bad, it was nowhere near as depressing as the finality of being sold to Elon Musk. Employees hugged and cried, unsure of how much longer they'd be working together.

Yue saw a colleague whom she recognized from the bike cage below the office. They both had the "mom bike" with an infant seat attached to the back. "How's it going?" Yue asked. The woman said she was resigning. "The chaos is too much," she explained.

Yue gave up on the festivities. She scooped up her kids; it was time to go home. As she walked toward the elevators, she passed Jay Sullivan, who was looking solemn. "It's done," he said.

The deal had closed.

Elon Musk's acquisition of Twitter cost roughly $44 billion. Musk committed more than $30 billion in equity financing, which included his existing $4 billion stake in Twitter, close to $20 billion in cash after selling part of his stake in Tesla, and approximately $7 billion that he'd secured from investors, including the venture capital firm Andreessen Horowitz and his billionaire friend Larry Ellison. Jack Dorsey also agreed to roll over his 2.4 percent stake in Twitter, saving Musk approximately $1 billion.

The rest of the funding came from major banks, which committed a whopping $13 billion in debt financing—a structure known as a leveraged buyout. Twitter borrowed money against its own assets, meaning the company—not Musk—would be on the hook if Twitter couldn't make its interest payments, which would amount to roughly $1.5 billion a year.

"He probably only got that much because he is Elon Musk, the richest person in the world, and banks wanted to be in his good graces so were willing to make aggressive loans to Twitter," says Bloomberg finance columnist Matt Levine.

The afternoon before the deal closed, Musk's team realized he was short about $500 million. Musk's team suggested Twitter's cash could fill the gap. The money was about to be theirs anyway. There was no precedent for such a suggestion, and Twitter's leadership team could not have complied if it wanted to, according to employees with knowledge of the situation. Eventually, Musk's team came up with the funding.

Twitter had been in a manageable financial situation prior to the acquisition. "Twitter was a mediocre business. Its cultural importance was always bigger than its income," Levine says, noting that its cash flow from operations was $1.3 billion in 2019, $993 million in 2020, and $632 million in 2021—"a bad trend, but still positive cash flow." GAAP net income was negative in 2020 and 2021. "So it wasn't doing great, but Musk's view that Twitter was in grave danger of imminent bankruptcy before he came along is not really shared by anyone else and seems mostly wrong," Levine explains. In fact, Musk's acquisition immediately put the company in a precarious financial position by saddling it with billions of dollars of debt and casting doubt on its primary income stream.

Twitter employees, who received a sizable portion of their compensation in the form of stock, would be paid out for vested shares at $54.20 within days of the deal close. Remaining shares would continue to vest on a quarterly basis. In the first half of 2022, Twitter recognized $459.5 million in employee stock payments, according to Lora Kolodny at CNBC.

The details of the deal had to hurt. Not only was Musk buying a company he wasn't sure he wanted, since 2021 the new CEO's net worth had fallen nearly 35 percent, from its peak of $320.3 billion on November 4, 2021, to $209.4 billion by the end of October 2022. Ironically, Musk's decision to sell $31 billion worth of Tesla stock to fund his part of the deal might have been partly to blame, as it looked to some like he'd lost faith in his car company.

Now that he owned Twitter, Musk's first order of business would be stripping away costs as quickly as possible.

Yoel Roth was wrapping up work in a conference room when he got a Slack message telling him Musk wanted to meet. He was nervous. He figured he was about to get fired. Everyone knew that Musk hated his boss, Vijaya Gadde, and thought "trust and safety" meant censorship.

Roth forced himself to smile as he walked past employees who were trying to enjoy the festivities with their kids. He didn't want to freak anyone out. When he crossed the sky bridge and arrived at the second floor of the building, Yoni Ramon, a Tesla security engineer, stopped him. "How do I get access to Twitter's internal content moderation systems?" Ramon asked without introduction.

Roth blinked. "You don't," he said. "That's not going to happen." He explained that Twitter was operating under an FTC consent decree, and accessing the company's internal systems was highly sensitive, something only employees could do after they'd signed the consent agreement and gotten proper training. The company was already under a microscope due to Mudge's whistleblower complaint. It needed to tread very carefully.

Ramon explained that Musk was worried an employee was going to try to sabotage Twitter. Roth suggested a solution: restricting broad access to Twitter's internal systems while keeping it enabled for critical content moderation functions. "OK, you're going to tell that to Elon," Ramon said.

Moments later, Roth sat face-to-face with Musk, walking him through Twitter's internal systems. "I showed him his own account in Twitter's enforcement tools, and I explained what the basic capabilities are," Roth later told Casey Newton. "And then I made a recommendation on what I thought Twitter should do to prevent insider misuse during the corporate transition."

Roth told Musk that Twitter needed to maintain its ability to moderate in Brazil. A contentious election between the sitting president, Jair Bolsonaro, and his leftist opponent, Luiz Inácio Lula da Silva, was just three days away. Musk agreed—he was in a conciliatory mood. "Bolsonaro and Lula are very dangerous," he said. "We need to protect against that."

The meeting had gone far better than Roth had expected. He was stunned at how reasonable Musk seemed face-to-face, compared to the bombastic character he played on Twitter. "I came into that meeting expecting him to fire me, and instead, he jumped ahead of me to say that he was sensitive to the risks of offline violence in the Brazilian election and wanted to make sure that we didn't mess with Twitter's content-moderation capabilities," Roth told Newton. "It was like a dream come true."

That night, Musk started clearing out the executive ranks. He fired Parag Agrawal; Ned Segal, the CFO; Vijaya Gadde, the chief legal officer; and Sean Edgett, the general counsel. Musk claimed that he'd fired the executives for cause, meaning they'd committed some act of misconduct that precluded Musk from having to pay out their contracts and severance. In all likelihood, the only transgression they'd made was having those contracts in the first place. Edgett was the only one in the office when the emails came through. He was unceremoniously escorted out by security.

"The bird is freed," Musk tweeted at 8:49 p.m. Pacific time.

Employees lingered on Slack waiting to get a note from Musk directly. But nothing came.

CHAPTER 20

"We Truly Cobbled It Together"

Musk wasted no time launching his first project as CEO. On October 27, 2022, hours after the deal closed, he told the engineering team that he wanted to change what logged-out users saw when they visited Twitter.com. At the time, if a new Twitter user visited the site, they were greeted with a sign-up form encouraging them to create an account. The landing page told users what Twitter was but gave them very little sense of what it was like. Wouldn't it be more compelling to potential users if they glimpsed a feed of Twitter's best offerings rather than some stuffy marketing copy? It was a smart idea. The home page had been static for years. If anyone had ever proposed changing it before, the concept had never been put into action. Now, Musk wanted users to be redirected to the Explore page to see trending tweets and stories. It was a statement: he had ideas about how to make Twitter better, and he planned to implement them immediately.

As if to underline this point, Musk told the engineering team they needed to have the project done by 9 a.m. the next day. The team had less than twelve hours to make the change. The old Twitter, with its endless debates and brainstorming sessions, was dead. Under Musk, he would supply the ideas and the company would focus on execution.

As the team rushed to complete the project overnight, Twitter's chief privacy officer, Damien Kieran, urged employees to consider the legal implications of the change. At a minimum, Twitter needed to do a Data Protection Impact Assessment—a formal risk assessment associated with processing user data—to comply with laws in Europe.

Kieran had been working closely with Musk's engineers, who technically all reported to him. Before they could get Twitter laptops, they needed to sign the FTC consent order, and Kieran got them printouts to try to ensure this would happen. He seemed to be working against the tide, trying to get Musk to slow down, or at least pay attention to the regulations, just as Musk was trying to speed up.

The engineering team worked through the night, and as promised, the new home page went live the next day. Kieran threw together a DPIA as quickly as he could, according to an employee who was involved in the project. Typically, the assessment would've been conducted by Kieran's team. Given the speed at which everything was moving, that wasn't an option. "We truly cobbled it together," the employee said.

Musk seemed pleased.

CHAPTER 21

"Comedy Is Now Legal on Twitter"

On Friday, October 28, a crowd of reporters swarmed outside Twitter's San Francisco headquarters, eager to see what Musk had in store for his first official day. The sky was slightly overcast and Market Street was humming with the usual mix of commuters and tourists. Finally, two men appeared, wearing black backpacks and puffer jackets. They were carrying cardboard boxes. One held up Michelle Obama's memoir, *Becoming.*

They told reporters their names were Rahul Ligma and Daniel Johnson and claimed they were part of a team of engineers who'd been laid off. "It's happening," wrote CNBC's Deirdre Bosa dramatically on Twitter. "Entire team of data engineers let go. These are two of them #Twitter Takeover."

She didn't catch the joke—"ligma johnson" is a popular internet meme where "ligma" stands for "lick my." The men were pranksters. Incidentally, they were also friends with Randall Lin. They'd met on Twitter.

Twitter employees shared Bosa's tweet on Slack. "For whatever the rumors are worth," one staffer wrote. Roughly thirty people responded with a "😔" emoji, perhaps assuming the layoffs were real.

Musk got the joke immediately and invited the men to the office: "Ligma Johnson had it coming 🍆💦" he tweeted. "Comedy is now legal on Twitter." Bosa soon issued an on-air apology.

I later pinged "Daniel" to ask if he'd speak to me on Signal, an encrypted messaging app. "Signal is compromised and unreliable," he said. "Several of us have moved to sugma." It was a continuation of the Ligma joke.

"At least you stay on brand!" I responded.

He tweeted a screenshot of the conversation, including my phone number. Technically, this violated Twitter's policy against publishing private information, but I didn't bother asking the company to take it down.

CHAPTER 22

"Please Print"

That afternoon, an executive assistant pinged Twitter engineers on Slack. Musk wanted to meet with them all one-on-one to review their work. "Please print out 50 pages of code you've done in the last 30 days (if you haven't submitted code in the past 30 days, then you can go back up to 60 days)," she said. "Please be ready to show on your computer as well . . . Recency of code is important but also use a code that shows complexity of our code."

Yao Yue was amused. The idea of printing out and reviewing code was extraordinarily strange. Was Musk actually going to review fifty pages of code from every single Twitter engineer?

Yue printed out some Python, one of the easier programming languages, and waited. *Python is more at Musk's level,* she mused. She wasn't going to print out the complex initiative she was actually working on. "I'm not gonna explain the project I've spent ten years working on in a fraction of an hour competing with ten other people—I'm just not," she said.

Around her, engineers scurried around trying to find printers that worked. Pages of code were soon scattered around the desks. Then another message came through on Slack.

"UPDATE: Stop printing," it read. "Please be ready to show your recent code (within last 30-60 preferably) on your computer. If you have already printed, please shred in the bins on SF-Tenth. Thank you!"

Again, a flurry of activity unfolded, as engineers rushed around trying to find paper shredders.

Yue waited, unworried. Her code review was pushed back and then canceled. She still hadn't heard from Musk since the small-group meeting two days prior. "We didn't actually get to show our code to Elon," she recalls. "Which is a shame. I was very much looking forward to it."

CHAPTER 23

"I Bet My Emergency Is Bigger Than Yours!"

Later that day, a high-ranking member of Twitter's finance team, whom we'll call Claire, walked into a meeting with Musk to discuss the issue that had been on everyone's mind since well before the deal closed: the layoffs.

Claire had helped Agrawal prepare for layoffs earlier that year. She already had a list ready to go—it just needed to be updated. "We can cut 16 percent of the staff," she said. She knew that Musk wanted to cut more—upwards of 70 percent, if the *Washington Post* article was to be believed—but hoped she could convince him to start small since she'd already done the legwork for him.

To her surprise, Musk agreed. But there was a caveat: he wanted the list ready to go by the following evening. In less than five days, employees had a vesting cliff, and Musk owed them more than $200 million in stock. Claire worried he might not want to pay.

She left the meeting and turned to a colleague. "Either we leave and go to SFO or we have to do this RIF [reduction in force]," she said. "We could go to Hawaii! We could just leave!"

She was only half joking.

Claire texted a senior member of the HR team and asked her to come to a conference room. "We have an emergency," she said.

"I'm in another meeting that is also an emergency," the woman responded.

"I bet my emergency is bigger than yours!" Claire said.

"The entire leadership team, except Leslie Berland, just resigned."

"We need to figure out who's getting laid off by tomorrow," Claire responded.

Claire did not flee to Hawaii. Instead, the two leaders got to work.

The Goons began calling managers, gathering intel on who should keep their job. "Who's critical?" they asked. "Who's *technical*? Who are the best people on your team?" They told managers to stack rank their teams, listing employees in order of performance.

Amir Shevat, head of product on Twitter's development platform, tried to get more information. What metrics should they use? Were managers supposed to rank people by seniority? Or work ethic? Or productivity? What about impact?

"They said, 'We don't know. Elon wants a stack rank,'" Shevat recalls.

Shevat wasn't sure what was going to happen to his team now that Musk owned the company. In his estimation, they'd brought in $400 million in revenue in 2021, with more than a 90 percent profit margin, but it wasn't clear Musk supported their work.

Shevat joined Twitter in March 2021 after Twitter acquired his startup, looking to correct a nine-year-old misstep in how Twitter engaged with third-party developers. In 2012, Twitter had severely curtailed third-party-developer access to its application programming interface, or API, effectively stopping outside programmers from building apps on Twitter. In layman's terms, an API is a bridge that allows different applications to talk to one another. Limiting access to the bridge stunted innovation, and Dorsey later admitted it was a mistake. When Shevat came on board, he brought his seven-person team with him, and Twitter gave him license to hire fifty more engineers.

"We envisioned a world in which you could share your favorite song from Spotify and listen to it live with all your followers on Twitter," Shevat later wrote. "We wanted you to be able to share your donation to your favorite cause and get your followers to donate as well through an integrated, GoFundMe–style experience. We wanted you to play Wordle inside Twitter, not just share the results. We wanted you to be able to interact with developer-powered apps inside the Twitter user experience. That was just the beginning: We also envisioned a true decentralization of the Twitter timeline. We wanted to let developers create and share their own timelines."

Shevat's vision was in line with where the tech industry was headed. While companies like Apple operated as walled gardens, more social platforms were embracing decentralization, giving users control over their digital experiences.

Unfortunately for Shevat, the notion of opening Twitter's API was directly at odds with how Musk ran his other companies. Part of his playbook at Tesla and SpaceX involved owning nearly every aspect of the supply chain. If he could hire engineers to build a product in-house, why outsource it to another company? Outsourcing—which had been the go-to methodology at Twitter 1.0—meant less control. And Musk liked to have full control. Tesla famously ran its business on homegrown enterprise resource–planning software called WARP, rather than relying on systems from SAP and others, according to CNBC's Lora Kolodny.

As layoff rumors swirled, Shevat held a meeting with his staff to try to allay their concerns. Then the Goons banned large meetings. "Group meetings are no longer a thing," they said, according to Shevat. "If you do that, you risk getting fired." Some managers were even told to stop communicating with their employees on Slack.

Shevat got a message telling him that David Sacks wanted to meet. A video call was scheduled for 1 p.m. Shevat waited. The meeting was pushed back. He waited some more. Finally, at 8 p.m, Sacks logged on. Shevat walked him through his product road map, laying out his vision for what Twitter could build with third-party developers, but Sacks barely looked up from his phone. Shevat felt like he was back pitching to bored venture capitalists who had no interest in investing in his startup.

When the call ended, Shevat felt like he was going to cry. It wasn't just about his job. He'd brought his whole team over to Twitter, believing Dorsey's promise to invest in all they had to offer. Now, he felt like he'd led his crew into a death trap.

A small group of Twitter managers gathered in the San Francisco office the next day to figure out who to cut. Already, Twitter's headquarters felt different, as people looked around with fresh eyes. Would Musk see the coffee bar as proof that Twitter employees were pampered? Did the sunny patio look like it was built for slackers?

Managers were given lists of hundreds of employees and told to sort them as quickly as possible. Many didn't recognize the people they were sorting. The lists had minimal information, just page after page of names and titles.

The Goons knew the process was going to be messy. At one point, they told a high-ranking member of Twitter's HR team to add a form to the termination email, asking people to indicate whether they'd like to stay. Eventually, the Goons were convinced this was not a good look and let it go.

As the project marched toward its inevitable conclusion, Claire learned that Musk would be laying off far more than 16 percent of the staff. He estimated that Twitter was losing $4 million a day and wanted to drastically cut payroll costs. During one meeting, the CEO's adviser, David Sacks, said he planned to cut 50 percent of the sales organization. Claire suggested starting with 30 percent. "It was like horse trading," she said.

Over the weekend, the Goons directed managers to write down a sentence for every person they wanted to keep, explaining why that person was exceptional. The assignment needed to be done by Sunday morning. Those who didn't have a sentence written about them would be laid off with two months of severance. Managers scrambled to justify why employees who were pregnant, or undergoing cancer treatment, should keep their jobs. "It was like *Schindler's List,*" one executive remembers. "Nobody slept that entire night." Some managers asked their peers to put them on the layoff list, worried that if they resigned they wouldn't get severance.

Yoel Roth was up until four in the morning writing sentences for his entire team, roughly two hundred people in total. He stared anxiously at the Google Doc, waiting for someone to open it, but no one did. Eventually, the lists were all merged into a master list for Musk. Every manager had ranked employees differently, making the results incoherent.

"If I were to get that list, I would probably throw it in the garbage because it's completely useless," Shevat says.

CHAPTER 24

"Should I Post It?"

On Monday, October 31, Musk arrived in New York for a series of meetings with the company's largest advertisers. Already, brands across Europe, the Middle East, and Africa had paused advertising, putting more than $15 million in near-term revenue at risk, according to internal documents. Many more had reduced their ad spend by upwards of 90 percent. This was Musk's chance to stop the bleeding.

Over the weekend, Musk's reputation in the industry had gone from bad to worse after he promoted a baseless allegation about Paul Pelosi, the husband of House Speaker Nancy Pelosi. On October 28, a crazed intruder had broken into the Pelosis' home and beaten Paul with a hammer. Musk shared a story suggesting the attack was the result of a drunken fight between Pelosi and a male prostitute. "There is a tiny possibility there might be more to this story than meets the eye," he wrote. It was exactly the kind of dangerous misinformation that advertisers worried would proliferate on Twitter, and here was Twitter's CEO, spreading it himself.

Twitter's remaining leadership hoped that if the agency bigwigs could meet Musk face-to-face they would see that he wasn't as wild as the character he played on Twitter. It was a gamble, as they'd seen during the June all-hands meeting, when Musk droned on about aliens, but they had little to lose. Musk had promised them that he planned to make Twitter a top-tier advertising platform, and they wanted to believe him. If the company was going to maintain its revenue, it needed advertisers to believe him, too.

That afternoon, Bill Koenigsberg, CEO of Horizon Media, one of the largest independent advertising agencies in the world, arrived at the office, along with a few Horizon executives. Musk sat with them in a conference room on the seventh floor of the office. He'd brought his mother, Maye Musk, for reasons no one could discern, as well as Jason Calacanis, the "grande" jumper; JP Maheu, the VP of ad sales; and one other Twitter executive.

Koenigsberg said that his clients wanted to know whether Musk was going to unsuspend Donald Trump's account. Musk responded that he was getting this question from "everyone." Then he pulled out his phone to compose a tweet: "If I had a dollar for every time someone asked me if Trump is coming back on this platform, Twitter would be minting money!"

"Should I post it?" he asked. Maheu said no. Musk laughed. The rest of the meeting went on without a hitch. Musk could be charming when he wanted to. It wasn't until later that Maheu realized Musk had posted the tweet.

The next day, Maheu got a call from a member of the HR team. He'd been fired. Moments later, he was escorted out of the office by security. Leslie Berland, Twitter's CMO, was cut the same day.

Advertising executives were playing a risky game. They were wary about encouraging their clients to run ads on Twitter. But at the same time, they were eager to ingratiate themselves with Elon Musk. The CEO ran multiple major companies outside of Twitter. If he changed his mind about the ad industry, he had the potential to become a major client. Imagine the fortune that could be made off Tesla commercials.

After the meetings in New York, Mark Read, the CEO of WPP, another of the biggest global advertising firms, tried to schedule a call between Musk and some of his larger clients, including Unilever. He wanted the brands to hear from Musk directly, to decide for themselves whether they felt comfortable running ads on Twitter. When Read reached out to Maheu to set a time, he was surprised to find out that the ad executive no longer worked at Twitter. Read asked Musk if he should coordinate with Musk's personal assistant. Musk said he didn't have one and told Read to text him directly to schedule the call. It was unusual, to say the least, for an executive like Read to be handling meeting logistics. But that was how Musk liked to operate. If advertisers wanted to keep working with Twitter, it was how they'd have to operate, too.

CHAPTER 25

"#SleepWhereYouWork"

Back in San Francisco, engineering managers started pulling people aside to work on Musk's highest-priority project to date. The CEO wanted to overhaul Twitter's verification system. For the first time ever, verification would be part of an existing subscription service called Twitter Blue, allowing anyone to buy a blue check mark.

Unlike YouTube, which targeted its subscription service at viewers, Twitter Blue would be aimed at creators, the people who posted tweets. It was an odd choice, and one that did not show a clear understanding of Twitter's user base. In 2022, a Pew Research study found "roughly half of US adults who use Twitter (49%) post fewer than five tweets per month." In other words, most Twitter users were lurkers and had little need to be verified.

Twitter's verification system dated back to 2009, when the manager of the St. Louis Cardinals, angry over being impersonated, filed a lawsuit alleging trademark infringement and misappropriation of name and likeness. As a result, Twitter introduced the iconic blue check mark, which would signal to users that the account was real.

In theory, the blue check mark was supposed to be reserved for prominent people. But nearly any member of the mainstream media could get verified within twenty-four hours with the help of a savvy social media manager. Musk was suspicious of the mainstream press; in his mind the blue check gave journalists a false air of credibility.

"Twitter's current lords & peasants system for who has or doesn't have a blue checkmark is bullshit," he said.

Seemingly impressed by her forwardness the day he arrived in the office, Musk tapped Esther Crawford to lead the project. He gave her team an aggressive deadline. Musk wanted the new Twitter Blue to launch on November 7; otherwise, those working on it would be fired.

Manu Cornet, the cartoonist, pitched in to help his colleagues. The job was enormous. It wasn't just that the team needed to overhaul Twitter's subscription service. They also needed to plan for every possible scenario: a surge in sign-ups, a flurry of cancellations, a momentary loss of connection between Twitter and the payment platform. Almost immediately, Cornet caught the bug of excitement emanating from Crawford and the Twitter Blue team. "I wasn't really thinking about whether those changes were good for Twitter," he later wrote. "I just really liked being part of that community, working together on a deadline." One night he stayed at the office so late that he ended up sleeping on a couch.

Crawford's team toyed with different prices. At one point, they considered charging people $19.99 a month for the blue check. When the news leaked to *The Verge*, celebrities were outraged. Some argued that their tweets amounted to unpaid labor. While a platform like Fox News could plausibly say that pundits needed Fox more than Fox needed pundits, on Twitter, the balance of power lay squarely with celebrities. If prominent figures abandoned Twitter, the platform didn't have much left to offer.

"$20 a month to keep my blue check? Fuck that, they should pay me. If that gets instituted, I'm gone like Enron," tweeted author Stephen King to his 7.1 million followers.

"We need to pay the bills somehow!" Musk replied. "Twitter cannot rely entirely on advertisers. How about $8?"

King never responded. Shortly afterward, Musk officially set the price at $8 per month.

On November 2, Evan Jones, a Twitter product manager, tweeted a picture of Esther Crawford lying on the floor in a sleeping bag, wearing an eye mask. Like Cornet, she'd begun sleeping at the office. Crawford quote-tweeted the post. "When your team is pushing round the clock to make deadlines sometimes you #SleepWhereYouWork," she wrote. The photo instantly went viral. Musk's fans commended her, while some of Crawford's colleagues rolled their eyes.

The breakneck speed was unsettling to members of the trust and safety team, including Yoel Roth, who worried that tying verification to a subscription would embolden bad actors. His team put together a document outlining their biggest concerns, ranking the highest-risk items as "P0."

In the document, the following risks were designated as P0s: "Impersonation of world leaders, advertisers, brand partners, election officials, and other high-profile individuals." Then: "Motivated scammers/bad actors could be willing to pay . . . to leverage increased amplification to achieve their ends where their upside exceeds the cost." Finally: "Legacy verification provides a critical signal in enforcing impersonation rules, the loss of which is likely to lead to an increase in impersonation of high-profile accounts on Twitter."

The team sent the document to Crawford and to Musk's lawyer Alex Spiro. They even discussed it with Musk himself. But the CEO refused to delay the deadline. Improbably, he believed that the new Twitter Blue would help Twitter catch bad actors. The subscription required a credit card and phone number. Armed with that information, Twitter could identify and block spam accounts.

Yoel Roth doubted this would work. For one, most Blue sign-ups were expected to come through the App Store, and Apple was unlikely to hand over people's credit card information. For another, $8 wasn't a huge deterrent.

"New, compromised, or otherwise inauthentic accounts may pursue verification in order to carry out scams and spam," Roth told colleagues on Slack. "While the cost may deter many financially motivated spammers, [trust and safety] does not expect this to be a meaningful deterrent for sophisticated or persistent threat actors."

He hoped Musk knew something he didn't.

CHAPTER 26

"'Maniacal' Urgency"

Soon, the engineering team was divided into people working on one of Elon Musk's pet projects and those waiting around to be fired. The exceptions were people like Yao Yue and JP Doherty, whose day-to-day work was critical to keeping Twitter online. Musk's product ideas weren't bad, but they were all over the map. In addition to relaunching Twitter Blue, he was exploring a payments platform, long-form video, long-form tweets, and encrypted direct messages.

Behnam Rezaei, an engineering leader, tried to reassure employees who hadn't been chosen to work on one of Musk's special projects that they weren't about to lose their jobs.

Rezaei encouraged people to focus on "shipping and delivering" rather than strategy and planning. "If you want to be in 'special' group this week, code and ship 5x as before," he wrote on Slack.

The mandate fit perfectly into Randall Lin's skill set, which was honed toward getting things done. "He needed to find a group of people that were as execution focused as he was," a former manager tells me. With Musk at the helm, he finally had.

Lin didn't wait to get pulled onto one of Musk's special projects. He finagled his way in. The teams were stretched thin, scrambling to meet Musk's tight deadlines. Lin started arriving at the office early in the morning and leaving well after dinner, lending a hand wherever he could.

On Slack, an engineering director urged Lin's colleagues to show a similar level of initiative. "I suggest thinking about ideas for cool product that you wish you were building at Twitter," the director said on Slack. "Can you get a demo out quick? Can you work with 'maniacal' urgency on something you would love to see in the product? If so, please go ahead, and share a demo directly with Elon. At best: you will get some feedback. You may be asked to ship it asap. At worst, you will be asked to stop and work on something else. Even in this case, at least you worked on something you love . . . Please don't wait for an opportunity to be handed to you," the director added.

No one knew when the layoffs were coming. But it seemed clear that if employees wanted to keep their jobs, they needed to work harder and faster than ever before.

CHAPTER 27

"Uncertain Times"

As fears about layoffs mounted, Manu Cornet decided to share a simple browser extension he'd built so his colleagues could download their emails, in case they found themselves unexpectedly back on the job market. He might be working on Musk's highest-priority project, but his loyalty lay squarely with his coworkers.

Cornet's years working on Gmail had made him confident about what was technically allowed under a corporate "email retention policy." The browser extension, he reasoned, did not inherently violate any rules so long as people only downloaded comments about their work, not the content of the work itself. "In these uncertain times, here's a newly published Chrome extension that makes the process of downloading Gmail conversations for safe-keeping a little easier," he wrote on Slack.

On November 1, 2022, Cornet was in a meeting about launching encrypted direct messages when his video went dark. He tried to check his work email but saw that he'd been signed out. He thought, briefly, that his computer had been hacked. Then he got a call from HR. Cornet had been fired—one of the first employees outside the executive ranks to get axed. Cornet was told his recent behavior had violated "multiple policies."

Cornet never found out what that behavior was, though he hypothesized that it was the browser extension. "I was foolishly prepared to give

Musk the benefit of the doubt," Cornet said. Twitter failed to pay him the full severance package he'd been promised.

Eventually, Cornet became the lead plaintiff in a class-action lawsuit, *Cornet v. Twitter*, accusing the company of violating both state and federal labor laws.

CHAPTER 28

"Your Role at Twitter"

Employees braced for layoffs, but the November 1 deadline arrived and, to their surprise, Twitter paid out their vested stock. Soon after, JP Doherty was sitting in front of his computer when he saw a message from a man named Christopher Stanley. Musk wanted to lock down Twitter's code base, blocking engineers from shipping new changes, Stanley said.

Change freezes weren't unheard of at Twitter. In advance of a major event—the World Cup, the Super Bowl—the engineering team would sometimes freeze the code base so a new bug didn't disrupt the system. But Doherty wasn't about to lock down the code base for someone he'd never even met. "I'm sorry, who are you?" he asked.

Then he pinged a few higher-ups to see what he should do. They told him that Stanley, who worked in Information Security at SpaceX, was part of Musk's inner circle. "If he says we have to do this, we have to do this," one explained.

Doherty did as he was told. He knew what this meant. Without a big event on the horizon, there was only one reason for the work stoppage. The layoffs were coming.

At 5:21 p.m. Pacific time on November 3, dozens of employees were hanging out at The Lodge, a pair of real Montana hunting cabins that Twitter's architects had purchased off Craigslist and installed in the office. Then all their phones started going off.

"By 9AM PST tomorrow, everyone will receive an individual email with the subject line: Your Role at Twitter," the email read. "To help ensure the safety of each employee as well as Twitter systems and customer data, our offices will be temporarily closed and all badge access will be suspended." The memo was signed, simply, "Twitter."

Before employees could process what was happening, the security team swept through The Lodge and told everyone to clear out.

"From 'Twitter' looool what fucking cowards," one employee texted me. "Your people *are* Twitter, you shits."

After months of uncertainty, many employees were relieved to know the end was in sight. Across the globe, tweeps converged on the #social-watercooler Slack channel to say goodbye. Hundreds posted the salute emoji and a blue heart, the unofficial symbols of Twitter 1.0.

Brian Waismeyer, a data scientist on the health team, posted a final message to Twitter's leadership.

"I think you could have done better than a vacuum followed by an email signed from 'Twitter,'" he wrote on Slack.

"News articles aren't comms. Tweets from an account associated with half-baked rants, copy pasted memes, and the occasional misinfo aren't comms," he wrote, referring to Musk's account. "Secondhand internal sharing and employee sleuthing aren't comms . . . I also hope the failure of this past week hangs heavy on you to remind you to do better."

Waismeyer was laid off at 12:34 a.m. Pacific time.

On Friday morning, Twitter's staff woke bleary-eyed to assess the extent of the damage. Musk had laid off about thirty-seven hundred people—roughly half the company—overnight. Those who still had access frantically scrolled through Slack to figure out who else remained.

Randall Lin wasn't a monster. He felt bad for those who'd been cut. But he thought Musk had made the right call. "We were bloated, we were really fatty," he says. Layoffs were necessary no matter who was in charge.

Elsewhere in the company, however, the problems were immediately apparent.

It wasn't just that Musk had cut the entire communications team and most of the marketing staff. That was expected. The new CEO had also made deep cuts to security and engineering, leaving Twitter without a single employee who knew how to run critical components of its tech stack.

Amir Shevat had run a roughly 140-person team responsible for the developer platform. As he had feared, almost everyone (himself included) was cut. The privacy and security teams lost roughly 75 percent of its technical program managers who were in charge of FTC compliance. Musk gutted the ethical AI team and wiped out the human rights organization. About 15 percent of the trust and safety team was let go. Ninety percent of the staff in India had been laid off, leaving Twitter with just a dozen employees managing one of its most important markets. The infrastructure team, which Yue was a part of, was cut in half. Even the brand safety team, responsible for making sure ads didn't run next to violative content, was slashed.

Only JP Doherty's team emerged more or less intact. Of the twenty-four people on Twitter Command Center (TCC), only Doherty's manager was cut. The Goons knew that if TCC was dismantled, the platform would go down in a matter of weeks.

The Goons soon asked Doherty if he'd like to become the director of TCC. It was a big promotion and came with a salary bump. But Doherty had no interest in profiting off the misfortune of his colleagues. He declined.

On Signal and Discord, remaining employees joined groups with names like "Twitter survivors" and "Grande jumpers" to strategize about what to do.

Musk had reluctantly placed laid-off employees on a two-month "nonworking notice period" to comply with US labor law, which requires that companies with more than one hundred full-time employees give workers sixty days' notice before conducting mass layoffs. The CEO would've preferred to just let everyone go immediately, but that wasn't possible, so instead he told security to cut everyone's access to the company's internal systems and promised to pay them until mid-January. (Later, Courtney McMillian, the former head of Twitter's benefits program, filed a class-action suit against X Corp. and Elon Musk, accusing both parties of failing to pay at least $500 million in severance to laid-off employees.)

The security team rushed to comply. But the process was so haphazard that while some people lost access immediately, others lingered in Twitter's systems for weeks. A Slack bot that paged engineers who were on call, meaning they were supposed to be available at any time of day or night in the event of a technical emergency, was accidentally deactivated, while a GitHub organization that contained "code samples from Twitter Developer Relations" was left with former employees as owners. "There are 519 members on GitHub for this org—most of whom do not work at Twitter anymore but some still have Owner permissions (including ability to rename, delete the org etc)," an employee wrote on Slack on December 7.

Later, the Goons realized they'd cut nearly every person who (stick with me here) knew the pass phrase to unlock the file that contained the root password for Twitter's servers, located at the company's three data centers. Without the root password, the company didn't have administrative access to its own machines. If an engineer needed to make a significant change, they'd have to wipe the servers and reboot them from factory settings. Luckily for them, JP Doherty eventually heard what was going on and gave them the pass phrase. He was the only person at Twitter who still had it.

Amir Shevat, former head of Twitter's developer platform, soon joined a throng of employees filing suits against Twitter for allegedly failing to give them adequate notice about the layoffs. Lisa Bloom, a celebrity attorney who advised Harvey Weinstein and victims of Fox News host Bill O'Reilly, agreed to represent them.

"Remember when Elon showed up when he first bought Twitter and he had that sink in his hands and said 'hey . . . I own Twitter now, let that sink in,'" Bloom said on TikTok, holding a sink of her own. "Well, it was an apt metaphor, wasn't it, because Elon has clogged up the system with hate crimes, he's treated workers like dirt, and he's sending the company down the drain." She pointed at the drain, her eyes widening for emphasis. "We are filing claim after claim, day in and day out. Hey, Elon, let *that* sink in."

If full-time employees had been treated poorly during the layoffs, contractors fared even worse.

On November 12, Twitter laid off roughly forty-four hundred of its fifty-five hundred contract workers, including content moderators who were responsible for identifying posts that violated Twitter's rules. Unlike full-time employees, the contractors did not find out about the layoffs via an email. They were simply locked out of Slack and Gmail.

"One of my contractors just got deactivated without notice in the middle of making critical changes to our child safety workflows," a Twitter manager wrote on Slack.

Remaining contractors soon found out that they would not get paid time off for Thanksgiving as they had in years past. "We have received confirmation that given the challenging business environment, Twitter will no longer move forward with holiday pay for flex contractors," an email from one of the subcontracting companies read. "We understand this doesn't come at the best time, but hope everyone can at least enjoy the long weekend and decompress with family and friends."

Alex Haagaard (they / them) @alexhaagaard - Oct 27, 2022
Since we're all talking about leaving twitter again can I bring back the idea of migrating to linkedin en masse and just totally destroying the professional vibe over there

Ben Rosen @ben_rosen - Oct 27, 2022
Welcome to **NEW TWITTER**
When you receive **420 LIKES** you will earn **1 MUSK BUCK**. When you earn **69 MUSK BUCKS** you can say **A SLUR**

Cullen Crawford @HelloCullen - Oct 28, 2022
maybe MAYBE ill stick around. i guess if elon musk bought my toilet I wouldn't stop shitting in it

@simplylay - Oct 30, 2022
The scariest thing about Halloween is that rent due at midnight

Katie Notopoulos @katienotopoulos - Nov 1, 2022
People are like "ugh I'm leaving Twitter" but I just saw a viral thread where someone argued it's classist to know how to boil an egg. You can't put a price on this, you'll have to drag me out

pathological supply avoidance @griph - Nov 2, 2022
for moral convenience we will be breaking down all human behavior into two categories: Privileges and Traumas. Privileges will be bad and Traumas will be good. this will make discourse far more efficient allowing you more time to divest yourself of Privileges and acquire Traumas.

Alex Heath @alexeheath - Nov 3, 2022
"Sacks" is really an incredible last name for someone helping to plan mass layoffs

Chris Bakke @ChrisJBakke - Nov 4, 2022
The layoff email from Twitter is the first email in history that should have been a meeting.

CHAPTER 29

"Weak, Lazy, Unmotivated"

Layoff survivors fell into two camps: those who still had their jobs but were now actively looking to leave, and those who still had their jobs and desperately needed the company to allow them to rehire some of the employees who'd been let go in order to actually *do* those jobs.

Two of the biggest areas of concern were brand safety and compliance. Twitter needed to make sure that ads didn't run next to violative content. Even more critical—it had to comply with the FTC consent order. The layoffs had severely impacted teams charged with protecting user data, according to the Department of Justice.

On Saturday, November 5, Musk's lieutenants told a small group of managers they could rehire some of the people who'd been cut. They had just twenty-four hours to submit a list of employees who should now be offered their jobs back. Only exceedingly high performers were allowed.

"Sorry to @- everybody on the weekend but I wanted to pass along that we have the opportunity to ask folks that were [laid] off if they will come back. I need to put together names and rationales by 4 PM PST on Sunday," wrote an engineering manager on Slack, asking his colleagues for any and all nominations.

Luke Simon, the engineering director who managed Twitter's ads stack, was one of the managers who needed to do some rehiring. He said he was seeing "silent quitting" that made him concerned about the people who were still working at Twitter. During the layoffs, he'd recommended deep cuts to the brand safety team and ad reviews team, which sat in the revenue organization called Goldbird. Now he told colleagues that he needed to bring some people back or he wouldn't be able to merge the ad review team with the health organization, as he'd planned.

"These people are not in the top 45% of Goldbird eng, but they are critical to transferring Ad Review to Health, and so I want to bring them back to Twitter," he told colleagues in a private Slack channel.

"I'm supportive here—Brand Safety is an immediate concern, we don't have time to rebuild expertise from the ground up," Murph Finnicum, a product engineering leader, responded.

"This is going to be the challenge," Simon said. "The engineers I am bringing back are weak, lazy, unmotivated, and they may even be against an Elon Twitter."

He added: "They were cut for a reason. So we need to think of these people as just needing to be around until the knowledge transition is completed."

Ella Irwin, the VP of product on the health team, said she'd discussed the situation "several times" with Musk's lieutenants and he was adamant that low performers couldn't come back. "We have multiple similar crisis situations," she said. "They were a hard no."

The conversation left one colleague confused about how to move forward. Did they want to hire people back, or did they not?

"Is everyone in this thread support[ive] of this approach of trying to bring these people back?" a product engineering director asked. "Bringing back folks that are a known wrong fit and/or don't want to come back and/or will be managed out quickly sounds suboptimal. Let me know if I'm missing something."

A senior manager on the strategy and operations team suggested a middle ground. Could they hire people back as consultants, there to transfer knowledge about the tech stack, but little more?

Irwin didn't agree with this approach. "I would not recommend bringing people who are the wrong fit back unless we think we made the wrong judgment call about their skills," she said.

Finnicum agreed. "We're going to have to be comfortable with the reduced execution on brand safety meanwhile," he said. "I don't have great perspective, but it seems like that's gone from a 'what-if' to a real business risk."

"Same with regulatory compliance," Irwin acknowledged. "Literally the big fire that is already burning."

The team decided to let Musk make the call. "These were conscious decisions that were made with the understanding that we would be able to manage through the head count gap, I believe," Irwin said. "All the discussions I was part of referenced dealing with these big risks by pulling engineers from wherever we had [them] in the company."

She added: "And of course the clear direction from his team [was] that we were not to keep any low or even slightly below average performers irrespective of the business risk."

They added these insights to a document they were preparing for the meeting, under a section called "fires."

"Yeah, Brand Safety is a clear fire and we can highlight the lack of expertise there rather [than] trying to judge ourselves," Finnicum added.

Later, a screenshot of Simon's comment regarding the "weak, lazy, unmotivated" engineers leaked on the anonymous employee forum Blind. Employees were outraged. Some tagged Simon on Slack, demanding answers. The screenshot eventually spread to Twitter, where it, too, went viral.

Shortly after, Simon held an all-hands meeting to address the leak. He said that he was sleep deprived and had been working twelve-hour days. But he didn't apologize. Instead, he lashed out at the leaker, saying those who "leaked private information were breaking the law, that they would be found and held legally liable."

CHAPTER 30

"New Blue . . . Coming Soon!"

Twitter's new subscription service was set to go live on November 7, 2022—the day before the US midterm elections. On Slack, employees asked pointed questions about why Twitter was "making such a risky change before elections, which has the potential of causing election interference."

Musk hadn't adopted the recommendations from trust and safety, but Yoel Roth was still hopeful that the CEO could be convinced to change his mind, at least about the timing of the launch. In early October, Twitter had temporarily locked Ye, the rapper formerly known as Kanye West, out of his account over an antisemitic post. Twitter removed the restriction on October 29, and West continued posting. In early November, Roth sent a particularly offensive Ye tweet to Musk. He told the CEO that the tweet was likely to inflame advertisers.

Based on the number of strikes Ye had accumulated on his account, the comment warranted a seven-day suspension. But Musk told Roth to ban the rapper outright. "He immediately understood that it was a test case and taking decisive action now could be used to recruit more advertisers," Roth said.

Finally, with three days to go until the midterms, Musk relented, pushing back the launch date to November 9, the day *after* the election. Employees who'd been worried he was purposefully trying to mess with the vote were relieved. "OK, this isn't nefarious, it's just gross incompetence," one told me.

The information about the delay didn't make it to the App Store. Over the weekend, Twitter users noticed that the new app release notes announced Twitter Blue was already live, ahead of the original deadline.

"Power to the people," the update read. "Your account will get a blue checkmark, just like the celebrities, companies, and politicians you already follow." (Jason Calacanis wrote the copy, which was panned for sounding like a phishing scam.)

Crawford tried to clear up the confusion on her personal Twitter account. "The new Blue isn't live yet—the sprint to our launch continues but some folks may see us making updates because we are testing and pushing changes in real-time. The Twitter team is legendary," she tweeted, adding a salute emoji. "New Blue . . . coming soon!"

On November 8, in an attempt to reassure brands that they'd still be able to mark their accounts as official, Crawford said the company would add an "official" label to some "select accounts."

The irony was immediately apparent. Crawford and Musk had just recreated the same "lords and peasants" system they'd aimed to dismantle, except with "official" labels instead of blue check marks. When a prominent creator pointed this out to Musk, he immediately overruled Crawford. "I just killed it," Musk tweeted. "Blue check will be the great leveler."

Jason Calacanis had gone from enthusiastically texting Musk his Twitter Blue ideas to sharing them, stream-of-consciousness style, on Twitter. "I predict that over time the Twitter Blue offering will have so many fun, helpful, & delightful features that people will feel like it is way too cheap at $8," he wrote on November 3. "The initial offering is the *start* of an obsessive rollout of fantastic features you will get real value from."

Employees had seen Musk be responsive to ideas floated by his Twitter followers. Complaints and suggestions from high-profile users could quickly turn into engineering projects.

But Musk didn't like what he was seeing from Calacanis. The angel investor was sitting in a conference room at Twitter's San Francisco headquarters when the call came down: Calacanis needed to stop tweeting about Twitter Blue. Musk didn't want him acting like he was leading the project.

"To be clear, Elon is the product manager and CEO," Calacanis tweeted, seeming apologetic. "As a power user (and that's all I am!) I'm really excited."

Verified Twitter users did not waste time trying to prove how confusing the new Twitter Blue was going to be. "I am a freedom of speech absolutist and I eat doody for breakfast every day," wrote comedian Sarah Silverman, after changing her display name to "Elon Musk."

Soon, Musk issued a new mandate: "Going forward, any handles engaging in impersonation without clearly specifying 'parody' will be permanently suspended," he said.

Sasha Solomon, a software engineer in Portland, Oregon, who worked on Twitter's core services team, was getting increasingly frustrated with Musk's hypocrisy. He said he was committed to free speech but seemed to be changing the rules to punish his detractors. Solomon quote-tweeted Musk's announcement. "full legal names only," she wrote. "plz submit your legal documents if you want to change your name." Then she added: "for example my full legal name is 'sach @ the hellsite' but if i wanted to change my twitter name to 'sach @ the combination hellsite dumpster fire' i'd need to submit my proof of legal name change."

On Sunday, November 6, actor Valerie Bertinelli decided to try her hand at parody. "The blue checkmark simply meant your identity was verified," she tweeted. "Scammers would have a harder time impersonating you. That no longer applies. Good luck out there!" Then she changed her display name to Elon Musk. She started tweeting as him, using hashtags like #VoteBlueToProtectYourRights. Within hours, her name was trending on the platform. "Okey-dokey I've had my fun and I think I made my point," she wrote Sunday night before changing her name back to her own.

Comedian Kathy Griffin saw Bertinelli's final "Musk" tweet. *I'll take it from here, Val,* she thought. She changed her name and profile picture to match the new CEO.

Griffin, an outspoken Democrat, wasn't a fan of Musk. After he'd purchased Twitter, she'd sold her Tesla stock. *This guy shouldn't be running a company,* she thought. *Like, any company, but certainly not an important company.*

As Musk, Griffin urged her two million followers to vote Democrat in the midterm elections. "After much spirited discussion with the females in my life, I've decided that voting blue for their choice is only right (They're also sexy females, btw.)," she wrote.

Hours later, Griffin noticed that her blue check mark was gone. Then her profile picture turned into a gray egg. Her account had been permanently banned.

"BREAKING: @KathyGriffin has been permanently suspended from Twitter for impersonating @ElonMusk," tweeted right-wing pundit Benny Johnson.

"Actually, she was suspended for impersonating a comedian," Musk responded.

"But if she really wants her account back, she can have it."

"For $8."

On November 10, Sean Morrow was sitting at home in Queens trying to ignore the piles of dirty laundry surrounding him when he saw a tweet that made him do a double take. It was Mario, the Nintendo character, giving the middle finger, and it had been posted by Nintendo of America, a verified account with the handle @nlntendoofus. Morrow kept scrolling. A verified LeBron James said "I am officially requesting a trade." By the time Morrow saw a verified George W. Bush tweeting, "I miss killing Iraqis," the progressive journalist knew exactly what was going on. Musk's new verification system had launched. It was mayhem. It was also, to Morrow, an opportunity.

Morrow dug up an old parody account he'd used when Joe Manchin was running for Senate in West Virginia. He paid $8 to get it verified. He considered doing something silly—pretending to be Subway and saying the footlong sandwich was now thirteen inches long—but then he had a better idea for how to use his newfound power.

Morrow googled "top insulin producers"; the name Eli Lilly popped up. He changed his handle to @EliLillyandCo, then copied the avatar photo used by the real Eli Lilly account. In his bio, he wrote the word *parody* multiple times, in compliance with Twitter's rules around satire.

Then he sent the tweet. "We are excited to announce insulin is free now." It was just corporate enough to sound believable.

About eight million Americans require daily insulin injections to help treat diabetes—a medication that could cost more than a thousand dollars a month. Morrow's tweet was soon shared more than three thousand times and sparked a crisis inside the pharmaceutical giant. Eli Lilly executives contacted Twitter to demand the post be taken down. It took six hours for Twitter to do so. In the meantime, the real Eli Lilly account tweeted that insulin was not, in fact, free. The company's stock dropped 4.37 percent. The following day, Eli Lilly paused all its advertising campaigns on Twitter, potentially costing the tech giant millions of dollars. Months later, Eli Lilly dropped the price of insulin.

When I spoke to Morrow, he seemed delighted at this near-perfect outcome. "They're both examples of corporate greed," he said about Twitter and Eli Lilly. "Elon was completely misunderstanding what verification was for. It's not a status symbol. It's in the name—it's verification."

Yoel Roth was too busy banning bad actors to say "I told you so." The trust and safety team was far from perfect, but it was damn good at putting out fires. "Everybody just moved into crisis response mode," he recalls. "Occasionally there would be a slight undercurrent in Gchat being, like, 'Wow, isn't this exactly what we predicted?'"

As major brands voiced outrage at being impersonated, the brand safety team sent around an FAQ to address their concerns. "Since Twitter Blue's launch, I've noticed blue check accounts impersonating people (LeBron James) and brands (Nintendo) on the platform. What are you doing to stop this?" it read.

"These examples have been extremely rare, but we are suspending the accounts quickly (and not refunding subscriptions, which reduces the incentive to re-offend)," the response said, in an attempt to downplay what was happening. "We anticipated early efforts like this from bad actors, and we are adapting dynamically to prevent and detect them. Further, as Elon tweeted, accounts engaged in parody impersonation must include 'parody' in their display name, not just in their bio." (This seemed like a direct response to Morrow's Eli Lilly account.)

The final question in the FAQ document read: "Do the same rules apply to Elon as to everyone else on Twitter?"

The company's simple, improbable answer: "Yes."

The afternoon Twitter Blue launched, Musk sat in a conference room in Twitter's San Francisco headquarters to host a Twitter Space with Robin Wheeler, head of advertising and sales, and Yoel Roth. The topic of the conversation was Twitter's commitment to advertisers.

Wheeler had been at Twitter for more than a decade and had deep ties to the marketing industry. "I am excited to be here and joined by . . . Elon

Musk, our CEO, chief twit, chief complaint officer, and what else are you calling yourself today?" she said on the call, sounding anything but excited.

Musk let out a nasally laugh. "Um, well, I'm a complaint hotline . . ." he said.

If the childish response irked Wheeler, she didn't show it. Without missing a beat, she promised marketing executives that Twitter's commitment to the advertising industry hadn't changed.

Roth, in his office, was having technical difficulties joining the Twitter Space. He was using an employee-only version of the app, and it kept crashing, leading him to download the consumer-facing version.

Finally, Roth jumped in and did his part, explaining that Twitter was actually *safer* under Musk than it had been previously. After the deal went through, the platform experienced a noticeable spike in hate speech, and Musk had allowed Roth's team to censor this speech more aggressively than ever before.

Prior to the acquisition, the trust and safety team viewed this speech differently depending on who was reporting it. For example, if the person reporting the tweet was the target of the speech, it was considered a first person report, and the tweet was taken down or de-amplified immediately, as long as it contained a single slur. If the person reporting the tweet was not the target, however, the report was considered a "bystander report," and the tweet needed to contain two slurs in order to be acted upon. After Musk took over, he expedited the project to detect and de-amplify hate speech, regardless of who reported it.

The trust and safety team had been heartened by this work. But Twitter Blue reversed much of that goodwill. "We're doing the right thing, but we're doing the right thing for all the wrong reasons," said a member of the team. "We're doing it to save face, not because he cares about the fundamental impact of hate speech."

Now, on the call, Roth did his best to defend Twitter Blue, despite his growing misgivings.

Roth desperately wanted to believe that Musk had a plan. If Musk was right and Twitter could identify verified scammers by somehow getting their credit card numbers from Apple, it would help the trust and safety team do its job.

Before Roth could think about this further, Wheeler turned the mic over to David Cohen, CEO of the Interactive Ad Bureau, one of the governing bodies of the ad industry. "We are rooting for you and for Twitter," Cohen said. He asked Musk how advertisers should think about the coexistence of his personal brand and Twitter's corporate brand.

"If I say that Twitter is doing something, I mean Twitter, and if I say 'I' then I mean me," Musk said. "If there's any confusion about the two, I would just ask me on Twitter."

On November 11, less than forty-eight hours after launch, Twitter paused all Twitter Blue sign-ups. Employees shared the update on Slack. "In order to prevent people from subscribing to Blue and to help address impersonation issues," the company had rolled out a number of product changes, a staffer wrote. Twitter had disabled the "stock keeping unit" for Twitter Blue subscriptions on the App Store, so iOS users couldn't sign up for the service. It also "hid the entry point to Twitter Blue." Existing Twitter Blue subscribers would still have access to Blue features, but no new subscribers could join—at least for now. In the two days since it had been live, fewer than sixty-one thousand users had subscribed, meaning Twitter Blue had brought in no more than $488,000, according to one analysis.

CHAPTER 31

"The Economic Picture Ahead"

On Wednesday, November 9, thirteen days after the takeover was complete, Musk sent his first company-wide email to Twitter employees. If anyone had hoped for the company's CEO to boost morale, they were sorely disappointed.

"Sorry that this is my first email to the whole company, but there is no way to sugarcoat the message," he wrote. "Frankly, the economic picture ahead is dire, especially for a company like ours that is so dependent on advertising in a challenging economic climate."

Musk laid out his plan. First, the company would go all in on Twitter Blue. "Without significant subscription revenue, there is a good chance Twitter will not survive the upcoming economic downturn. We need roughly half of our revenue to be subscription," he said.

Second, Twitter employees would have to work hard and be in the office every day. "Starting tomorrow (Thursday), everyone is required to be in the office for a minimum of 40 hours per week," he wrote. "Obviously, if you are physically unable to travel to an office or have a critical personal obligation, then your absence is understandable."

To some extent, employees had expected this announcement. Remote work wasn't allowed at Tesla and SpaceX. But the severity and speed of the new policy made it seem like a loyalty test. For the last two years under Dorsey and Agrawal, employees had operated under the assumption that they'd be able to work remotely forever.

In February 2022, before Musk had publicly expressed interest in buying Twitter, the company had filed a 10-K with the SEC saying that "even before the COVID-19 pandemic drove a shift to remote work, we recognized the need to evolve our workforce to achieve our purpose," and noting that "recent employee surveys show that the future of work at Twitter is hybrid, with a substantial majority of our workforce planning to work from home full time." Many employees had moved away from the Bay Area during the pandemic. They couldn't just move back overnight.

Lin was sympathetic to his colleagues who *couldn't* come in. But he suspected that many simply wouldn't. In his mind, Twitter's old office policy was a zero-interest-rate phenomenon. How could anyone expect to gain Musk's trust if they never showed up in person? Many weren't even on Slack past working hours!

"If I was bedridden, I'd still be available on Slack for questions," Lin grumbled.

But not everyone could prioritize work above all else. JP Doherty needed to be in Alameda, about fifteen miles from Twitter's San Francisco office, at 2 p.m. every weekday to transfer his son from the school van into his wheelchair, and then into the house. He told Musk that he'd need an exception. Part of him hoped that Musk would say no and he'd be forced to resign. But Musk—perhaps cognizant of how critical Doherty's team was to his success—said yes.

CHAPTER 32

"Elon Puts Rockets into Space, He's Not Afraid of the FTC"

The November 4 layoffs severely impaired Twitter's ability to comply with the FTC consent order. Now, the company was expected to lose even more staff as a result of Musk's return-to-office policy. Leaders across the company were panicking.

As part of the consent order, Twitter needed to have a designated owner for each of its privacy controls, which included database safeguards for sensitive user data and policies around information security. Unfortunately, Twitter no longer had anyone responsible for about 37 percent of Twitter's privacy program controls, according to Damien Kieran, then the chief privacy officer, per a deposition with the FTC. Lea Kissner, then the chief information security officer, said that half the controls in the information security program did not have a designated owner.

Kissner told the Department of Justice that the layoffs impaired Twitter's ability to "complete improvements in its data management, access, and deletion practices." They also said that "certain programmatic protections relating to product launch reviews, data access controls, and other ongoing security controls were effectively dismantled." Project Eraser, the initiative to honor a user's deletion request by automatically finding and removing the entirety of their data from Twitter's servers, was stuck in limbo.

On November 9, Kieran and Kissner resigned, along with Twitter's chief compliance officer, Marianne Fogarty. The group was comprised of the three remaining members of Twitter's Data Governance Committee, which was responsible for making sure the company's data practices were in compliance with the FTC consent order.

"I don't watch Game of Thrones. I certainly don't want to play it at work," Fogarty tweeted.

A senior member of Twitter's legal team broke the news of the high-level exits on Slack the following morning. "Everyone here should also know that our CISO, Chief Privacy Officer and Chief Compliance Officer ALL resigned last night. This news will be buried in the return-to-work drama. I believe that is intentional," she wrote.

She continued: "Over the last two weeks, Elon has shown that he cares only about recouping the losses he's incurring as a result of failing to get out of his binding obligation to buy Twitter. He chose to enter into that agreement! All of us are being put through this as a result of the choices he made.

"Elon has shown that his only priority with Twitter users is how to monetize them. I do not believe he cares about the human rights activists, the dissidents, our users in un-monetizable regions, and all the other users who have made Twitter the global town square you have all spent so long building, and we all love.

"I have heard Alex Spiro (current head of Legal) say that Elon is willing to take on a huge amount of risk in relation to this company and its users, because 'Elon puts rockets into space, he's not afraid of the FTC.'"

The lawyer urged her colleagues not to bow to Musk's return-to-office demands. "I do not, personally, believe that Twitter employees have an obligation to return to office. Certainly not on no notice (if at all)," she wrote.

The high-profile departures quickly made headlines. When Yoel Roth's sister heard the news, she sent him a text. "Don't be the guy left holding the bag," she warned.

CHAPTER 33

"Resignation Accepted"

Before Elon Musk bought Twitter, Yoel Roth wrote down a list of personal red lines that he wouldn't cross for the incoming CEO, which he described on *This American Life* to Casey Newton. He wouldn't break the law, or undermine an election, or "take arbitrary or unilateral content-moderation action."

But when he finally quit on November 10, 2022, it wasn't for anything he'd written on the list.

It had only been a day since Twitter Blue had launched. But the platform already felt like a mess. The platform was swamped with impersonators, and Musk wanted Roth's team to block them.

In the back of his mind, Roth wanted to believe that Musk had a secret deal with Apple. The CEO seemed certain that Twitter could identify bad actors using credit card information provided by the iPhone maker.

But now Esther Crawford was telling Roth something horrifying: Apple didn't give Twitter the phone numbers or credit card information of Twitter Blue subscribers. In fact, Apple didn't give Twitter *any* specific identifiers for individual users.

Roth tried to clarify. "If we catch someone for abusing the Blue badge, they can just keep using that same device and Apple ID to abuse for an additional $8 as they see fit?" he asked in the Blue Verified Slack channel. "That's contrary to what Elon had said and some of the operating assumptions here . . ."

An engineer responded to Roth earnestly, hoping to be helpful. "I'm no expert, but in the in-app purchase documentation, I don't see any Apple ID passed back to us." He confirmed that Roth's worst suspicions were true.

"Ok," Roth said.

"Happy hunting," the engineer replied.

Roth immediately called Musk and explained the situation.

Musk told him to call Apple and get the information. Roth tried to explain that Apple was never going to give Twitter private information about its users. Moreover, Twitter was already in hot water with the company, which had reached out that day to talk about impersonators and Twitter Blue. If Twitter didn't shape up, its next app release could be delayed. As he talked, Roth realized it was hopeless to argue with Musk. He moved on to a second pressing point.

To detect and ban verified impersonators, Twitter needed to redirect its remaining content moderators, many of whom were contractors, to work on the project. Those moderators needed to be thoroughly trained. It would take at least a couple days.

"They need to do it today," Musk said.

This time, Roth didn't bother pushing back. "OK, I'll get right on that," he said and hung up the phone.

Then he sent in his resignation. "I am no longer able to perform the responsibilities of my job and resign it as of today at 5:00 PM," he wrote.

"Behind Elon Musk, I was the most prominent representative of the company, period," he later told Casey Newton. "And I became aware that when Twitter Blue turned into the predictable hot mess that it was, that people would ask, why didn't the trust and safety team see this coming? Yoel, why are you so bad at your job?"

When Ella Irwin learned Roth had resigned, she texted him to ask what had happened. Roth told her the truth. It wasn't just one thing. There were the growing concerns over the FTC, the escalating fight with

Apple over Twitter's content moderation policies, and the absurd fiasco of Twitter Blue.

Irwin laid it on thick, telling Roth he was Twitter's only hope.

Roth got in his car—a leased Tesla—and drove home.

While Roth was driving, his colleagues in San Francisco were gathering in the Twitter cafeteria for a last-minute all-hands meeting with their new CEO. Roth had timed his exit strategically. He had no intention of ending his seven-year career at Twitter by being marched out the door by corporate security.

Musk walked in fifteen minutes late. Esther Crawford let out a cheer. An employee sitting next to her had to hold back a laugh.

Once Musk started talking, however, the mood in the room became solemn. The CEO made no secret of the fact that Twitter was running out of money.

Musk explained that bankruptcy was a looming possibility. "We just definitely need to bring in more cash than we spend," Musk said. "If we don't do that and there's a massive negative cash flow, then bankruptcy is not out of the question. That is a priority. We can't scale to one billion users and take massive losses along the way. That's not feasible."

Then he launched into a diatribe against remote work. The pushback he was getting seemed ridiculous. If people building cars and serving food couldn't work from home, why should the laptop class get to do so? "Let me be crystal clear," he said. "If people do not return to the office when they are able to return to the office, they cannot remain at the company. End of story . . . Basically, if you can show up in an office and you do not show up at the office, resignation accepted. End of story."

Employees posted a steady stream of commentary on Slack, as if trying to prove that Twitter's culture was still intact. "Somebody ask him about aliens!" one person joked. "This meeting could've been an email, yesterday's email [ending remote work] should've been a meeting," a colleague said. "Is [Elon Musk] a college freshman who forgot they had to do a public speaking assignment?" a third employee asked. "HOW ARE WE GOING TO COMPLY WITH THE FTC?" yet another message read.

Yao Yue watched the commentary, amused. She rarely attended all-hands meetings these days. She certainly wasn't going to start when she'd been given an hour's notice. When she saw her colleagues discussing Musk's comments about remote work, she knew her time had come to speak out.

"Don't resign, be fired. Seriously," she urged her colleagues on Slack. She posted a similar message on Twitter.

It wasn't about working from the office—Yue did that herself, every day. It was the principle of Musk asking people to resign for a policy he had just created. If Musk wanted to fire them, so be it. She didn't want her colleagues to do the CEO's dirty work for him. "I was acting in the spirit of the old company values, doing what I thought was right," Yue says.

Later that evening, Jared Birchall, the manager of Musk's family office, called Roth and asked him to reconsider his resignation. Roth declined. Birchall asked if Roth would talk to Musk, in an informal exit interview. When the CEO got Roth on the phone, Roth was careful to stick to his talking points. He'd left because "rapid and chaotic product development is setting fire to brand equity," he said. "Blue Verified was a misstep. We rushed it out the door, against extensive advice, and it failed in exactly the ways we predicted it would. That erodes my reputation and credibility."

Roth told Musk that he'd been pulled in too many directions. "Cut costs is our top priority," his talking points read. "But eliminate the impersonation immediately. But lay off more people."

Musk said he understood.

Keifer @DannyVegito—Nov 3, 2022
Elon Musk may be a libertarian but he's begging for 8 dollars with the intensity of a lifelong democratic senator

Lauren Dombrowski @callmekitto—Nov 6, 2022
being on Twitter right now is like playing the violin on the titanic except we are also making fun of the iceberg and the iceberg is getting genuinely mad

John Frankensteiner @JFrankensteiner—Nov 6, 2022
Every Elon post is like watching Joe Pesci enter the Home Alone house

Andrew Nadeau @TheAndrewNadeau—Nov 7, 2022
Everything happening on Twitter now is a lot easier to understand if you've ever had a younger sibling that invented a game and added a new rule every time they started losing.

sarah jeong @sarahjeong—Nov 9, 2022
every time the kind of wild bullshit that scares advertisers goes down on here, you fucking bet I'm opening up this hell app and refreshing the timeline repeatedly. eyes glazed over and unblinking as I inject brand-unsafe chaos straight into my veins 🥰🥰🥰

CHAPTER 34

"High Risk"

By the second week of November, the financial ramifications of Musk's decisions were starting to come into focus. Major brands, including T-Mobile and General Electric, paused advertising due to brand safety concerns. Dentsu, the largest ad agency in Japan, asked Twitter for a meeting. This was ominous.

Then GroupM, another huge worldwide advertising agency, updated Twitter's brand safety guidance to "high risk," meaning brands should only use the platform if they were comfortable with their posts potentially appearing next to harmful content. The move was a direct response to the high-profile departures in Twitter's privacy and security apparatus. "This does not necessarily mean that all brands will pause today, but it will be an influencing factor in how their customers approach Twitter in the coming days and weeks," a Twitter employee wrote on Slack. "They feel that Twitter's ability to scale and manage infractions at speed is uncertain at this time."

The agency said it would consider lowering the risk grade if Twitter could meet the following requirements:

- A "return to baseline levels of NSFW/toxic conversation"
- Fill senior roles on IT Security, Privacy, Trust and Safety
- "Establishment of internal checks & balances"

- Transparency around future plans that will affect user or brand safety, including changes to community guidelines and moderation policies
- A commitment to "effective content moderation," and ability to enforce the platform's rules

Employees on the sales team were at a loss. Ads made up 90 percent of Twitter's revenue. On Slack, they posted questions they were getting from advertisers. "Does your team have any guidance regarding the backlash aimed at healthcare brands who are choosing to continue to promote tweets under Elon Musk's leadership?" read one. "Question from AT&T re our reporting of hate speech tweets/imp[ressions]," began another. "Does our data only include the original tweet, or does our reporting include tweets + replies (i.e. ALL Data)?"

The question about how Twitter reported violative speech was tricky, as Musk claimed impressions on hate speech had declined since the takeover. In contrast, the Center for Countering Digital Hate (CCDH) found that hate speech had spiked.

"They found the n-word was used 30,546 times from November 18th to 24th, the week leading up to Musk's claims. That's 260% [more] than the weekly average for 2022. During that week, a slur for gay men rose 91%; a slur for transgender people rose 63%; a slur for Jews rose 12%; and a slur for Latinos rose 64%," reported *Gizmodo*, in an article that cited CCDH's research.

Advertisers weren't sure whom to trust. Roth might've claimed that Twitter was safer under Musk than it had been previously, but now he was gone and Twitter seemed more volatile than ever before. "A consistent theme is coming out of the meetings I'm having with clients," explained a Twitter employee on Slack. "They are asking if we work with a third party to verify what we are saying around decreased numbers of hate speech."

A colleague responded: "The heads of IPG Canada just asked the same thing. This will go a LONG way. Thank you!"

Musk didn't blame himself for the drop in advertising—he blamed CCDH. In 2023, Twitter (which by then had rebranded to X) sued the nonprofit, claiming it had embarked on a scare campaign to drive away advertisers in an attempt to stifle free speech.

"In direct response to CCDH's efforts, some companies have paused their advertising spend on X," the lawsuit claimed. It estimated the damages to be "at least" tens of millions of dollars.

CHAPTER 35

"How It Started/How It's Going"

In mid-November, Musk heard that Twitter was running slowly in countries like India because the app was batching remote procedure calls, or RPCs, inefficiently. If a user in India tried to refresh the app, it took about twenty seconds to load. It was unclear where Musk had heard this information, but he'd surrounded himself with people eager to tell him what he wanted to hear. "Btw, I'd like to apologize for Twitter being super slow in many countries," Musk tweeted. "App is doing >1000 poorly batched RPCs just to render a home timeline!"

Twitter engineers were annoyed. Twitter *was* a little slow in India—but it had nothing to do with how Twitter batched RPCs. Musk's tweet showed a profound misunderstanding of how the app worked. Eric Frohnhoefer, an engineer who worked on Twitter for Android, tweeted a levelheaded thread debunking Musk's claim. He pointed out that the Android app made less than twenty network calls to show the home timeline.

"The backend services typically use RPC to communicate," he told my colleague Casey Newton in an interview. "However, this is done within the data center over very fast network links. As some of my peers have pointed out, the >1000 RPC calls are unlikely to be the cause of performance issues."

Another engineer explained that the issue had to do with the physical distance between phones and data centers for Indian users, and the lower-powered phones that are popular in the country. "The fact that he's focusing on performance being worse in certain countries kind of shows that he doesn't know what he's talking about," the engineer said.

When Sasha Solomon, the thirty-four-year-old software engineer in Portland, saw Musk's tweet, she was furious. Solomon worked in Twitter's core services group and was intimately familiar with the company's digital infrastructure. "You did not just layoff almost all of infra and then make some sassy remark about how we do batching," she wrote in a quote tweet.

The following morning, at 1:44 a.m. Pacific time, Musk called an emergency meeting. He wanted to implement another change freeze. Exceptions would be granted only if there was an "urgent change that is needed to resolve an issue with a production service, including any changes reflecting hard promised deadlines for clients," and engineers got "approval from VP level and Elon explicitly stating that the change needs to be made," according to an internal email.

Employees had assumed that the layoffs would stop after November 4. But without a big external event to explain why Musk had frozen production changes, a new rumor started: a second round of layoffs was about to begin. There was no other way to explain why someone who insisted on cutthroat deadlines was now willing to grind work to a halt.

Musk was growing increasingly paranoid about the threat of internal sabotage. He told employees he was worried someone was going to *delete all the tweets*. Given how Twitter's architecture was set up, this threat seemed highly unlikely. But employees couldn't convince him otherwise.

In November, Musk had a meeting with a tweep in San Francisco, and the discussion grew heated. Few knew exactly what happened in that room, but there were rumors of physical contact. In the aftermath of the meeting, Musk seemed to grow more wary. By the time engineers started criticizing him on Slack and Twitter, he was out for blood.

The day after the RPC tweet, Solomon was sitting on the couch in her living room when she realized she'd been locked out of Slack. She tried her work email—also locked. Then she opened up her personal email. "We regret to inform you that your employment is terminated effective immediately," read a short note from "Twitter HR." And the usual, cryptic explanation: "Your recent behavior has violated company policy."

Across the company, more than a dozen engineers, including Eric Frohnhoefer, were fired. All had criticized Musk on Slack or Twitter. Clearly, Musk's commitment to free speech did not extend to his own workforce.

Yao Yue was lying in bed at 5 a.m. Pacific time the following morning when she found out that she, too, had been fired. What a way to start the day. She suspected she'd been cut because of the Slack and Twitter messages she'd posted, although she couldn't say for sure. The reasons were vague and referenced company policy, but never which one exactly.

In a previous life, Yue would have been devastated to lose her job, but now she felt only relief.

"After 12 amazing years and 3 weeks of chaos, I'm officially fired by Twitter," Yue tweeted. "Never expected I would have stayed this long, and never expected I would be this relieved to be gone. I have a lot of stories to tell. But to my fellow (ex-)tweeps- #LoveWhereYouWorked." She ended with the salute emoji.

The next day, she went to the office to retrieve her belongings: a few books and trinkets she'd gotten from colleagues over the years. Her badge had been deactivated so she went to the service elevator, where the security guards let her up. They'd known Yue for years and seemed to have more loyalty to her than the irascible new CEO.

After news about the targeted firings broke, I expected Musk to be contrite, still operating under the naive assumption that he cared about his reputation as a free speech absolutist. But Musk was far from apologetic. "I would like to apologize for firing these geniuses," he tweeted. "Their immense talent will no doubt be of great use elsewhere."

Solomon shared news about her firing on Twitter. She'd already become a figure in Musk's saga. She didn't see the point of hiding now. "lol just got fired for shitposting," she tweeted. "i said it before and i'll say it again

"kiss my ass elon 💋"

Chaya Raichik, an anti-LGBTQ crusader who posts under the username @libsoftiktok, screenshotted Solomon's remarks, first pushing back on Musk's understanding of RPCs and then telling her followers she'd been fired. "How it started / How it's going," Raichik wrote.

"A tragic case of adult onset Tourette's," Musk responded.

CHAPTER 36

"Extremely Hardcore"

The targeted firings of talented engineers tanked employee morale. Randall Lin worried the dour mood would impact people's work. He decided to fly to New York with his long-distance girlfriend, Alex, an OnlyFans star who lived in the Midwest, to check on two of his colleagues in person. "People were emotionally a wreck," Lin said. He paid for the flight and hotel himself.

Lin spent the week ping-ponging between late-night dinners with Alex and coworking marathons with his colleagues, where he did his best to give them a sense of direction and mission. Sure, the CEO was unpredictable, and yes, the recent firings had sucked. But those who remained had a chance to make history, building a super app alongside the CEO of SpaceX. It was worth it.

Then, on November 16, Lin woke up to see a new email from Musk with the subject line, "A Fork in the Road."

> Going forward, to build a breakthrough Twitter 2.0 and succeed in an increasingly competitive world, we will need to be extremely hardcore. This will mean working long hours at high intensity. Only exceptional performance will constitute a passing grade.
>
> Twitter will also be much more engineering-driven. Design and product management will still be very important and report

> to me, but those writing great code will constitute the majority of our team and have the greatest sway. At its heart, Twitter is a software and servers company, so I think this makes sense.
>
> If you are sure that you want to be part of the new Twitter, please click yes on the link below.
>
> Anyone who has not done so by 5 pm ET tomorrow (Thursday) will receive three months of severance.
>
> Whatever decision you make, thank you for your efforts to make Twitter successful.
>
> Elon

Musk had still not articulated a sophisticated vision for Twitter's business, but at least now he'd communicated an ultimatum about its culture. Twitter 2.0 would be technical- and engineering-focused. It would be, in Musk's words, "extremely hardcore." Get in or get out—and also, decide in the next thirty-six hours.

Lin's heart raced as he read the email. It was the moment he'd been waiting for. Since Musk had come on board, it felt like his colleagues had split into camps: the loyal few who trusted Musk as a leader and were willing to do whatever it took to build a new company, and those who were resistant to the new normal and were waiting around to be fired. Lin knew which camp he was in. Now Musk would know it, too.

He clicked yes, then called a handful of close colleagues. "This is an opportunity. If we lose our jobs, it's not that bad. What is severance going to get you, anyway? These are fun times to live in," he told them. "At least you'll have a story to tell."

Most were still on the fence. "We're being asked to keep a sinking ship afloat," one told him. By early Thursday, November 17, an internal poll showed that only about 25 percent of the software engineering organization planned to sign the hardcore email.

JP Doherty did not want to sign the email. But he knew he didn't have a choice. His son, Rhys, was scheduled to have strabismus surgery in January, correcting an eye issue that made it difficult for him to walk on his own. The procedure cost $10,000 out of pocket. Doherty discussed the decision with his wife, and while she wanted him to be able to quit, they both knew the kids needed his health insurance.

Still, Doherty didn't mind making the Goons sweat.

As the deadline approached, Doherty was eating lunch in the Twitter cafeteria when Sheen Austin, a Tesla engineer who was acting as the head of infrastructure, stopped to talk. "You can't do this to your team," Austin said. "They need you. Twitter needs you."

Doherty looked him squarely in the eye. "Why are you telling me this?" he asked. "Why isn't this coming from Elon?" He was annoyed that Musk's minions were trying to clean up their boss's mess. If Musk thought Doherty was so important, the CEO could tell him that himself.

Austin looked surprised. "You want to get in front of Elon?" he asked.

Doherty rolled his eyes. These sycophants just really didn't get it.

Nevertheless, Austin set up a meeting for later that afternoon between Musk, Doherty, and a few high-ranking members of the engineering team who hadn't yet signed the ultimatum.

One on-the-fence Twitter employee pulled up a presentation about Twitter's core values, including "Defend and respect the user's voice" and "Communicate fearlessly to build trust."

The employee said that it seemed fairly obvious Musk didn't intend to follow these values. Doherty was struck by the man's bravery.

But Musk brushed it off, launching into a monologue about the success he'd had at Tesla and SpaceX. "If you want to win, stick with me," he said.

No one said anything after that.

The group got up and walked out of the conference room. Austin hurried after them down the hall. "How'd that go?" he asked. "What did you think?"

An engineer walking next to Doherty gave an emphatic thumbs-down.

Doherty agreed with the engineer's assessment. But he didn't have the luxury of walking away. He'd taken the job to support his family; nothing about those priorities had changed. Two minutes before the deadline on November 17, Doherty signed the ultimatum.

Nearly one thousand of Doherty's colleagues used the hardcore email as their excuse to walk out the door. Twitter, which had seventy-five hundred employees in October when Musk walked into Twitter HQ with a sink, was down to just twenty-seven hundred full-time staffers. Musk had expected some developers to leave, but certainly not *that* many.

Once the deadline passed, Musk's lieutenants were left with a list of employees who'd opted into Twitter 2.0 and had to reverse engineer it to figure out who was leaving the company. In the meantime, hundreds of employees who presumably didn't believe in Musk's vision still had access to Twitter's internal systems.

Later that night, employees received an email telling them the offices were closed, effective immediately. Employees' badge access was suspended until the following Monday.

At 8:51 a.m. Pacific time, Musk sent another email. This one was just a single line: "Anyone who actually writes software, please report to the 10th floor at 2 pm today," he said. Employees were confused. Hadn't he just closed the office?

Twenty-five minutes later Musk added: "If you're working remotely, please email the request below nonetheless and I will try to speak to you via video. Only those who cannot physically get to Twitter HQ or have a family emergency are excused. These will be short, technical interviews that allow me to better understand the Twitter tech stack."

Eight minutes later he sent a third email: "If possible, I would appreciate it if you could fly to SF to be present in person. I will be at Twitter HQ until midnight and then back again tomorrow morning."

Lin told his girlfriend they had to cut the trip short and booked the next flight out of New York. She was annoyed, but she knew there was no convincing Lin to stay. They piled into an Uber and headed for the airport. On the way, Lin called the hotel to tell them he was canceling the rest of the trip.

Then Lin messaged his colleagues. "He's demanding we drop everything and just fly to San Francisco?" one colleague said, sounding aggrieved.

"He's not demanding. This is an opportunity," Lin said. He wondered where the line was between coercion and influence. "If you care, you'll be there."

Once Lin was on the plane he felt calm. He didn't have a plan for what he was going to do when he arrived at the office. Lin's favorite word kept going through his mind: "propinquity." Being there, being available, was enough. The next right move would reveal itself in time.

Lin landed in San Francisco at 6 p.m. and went straight to the office. He raced up to the tenth floor. The room was filled with engineers who'd been nervously sweating for hours. It smelled musty. Lin steered clear of the anxious tweeps and sat down next to a group of employees from Tesla, SpaceX, and Neuralink. James Musk, Elon's cousin, was sitting cross-legged on the floor. "It was exciting," Lin said. These engineers understood what it meant to be extremely hardcore.

"With Elon, you just have to know that no matter how good things are going, every day could be your last," one of the engineers advised Lin.

Lin overheard a Twitter engineer joking about Twitter's lax security policies. "Who cares about security?" he said, presumably trying to be funny. Lin cringed.

One by one, engineers got up to do code reviews with Musk and his lieutenants. The sessions were short, often just a few minutes each. To some, it seemed like just another excuse to fire people. "The 'code reviews' were a clear pretext to attempt additional 'for cause' firings;" an employee lawsuit, Arnold v. X Corp., later alleged. This wasn't an opportunity for Musk and the Goons to understand the tech stack; it was a trap. "The 'reviewers' lacked the context to meaningfully evaluate the code, and the reviews were completed in an amount of time that was clearly insufficient for any good faith approach to the task."

After a few hours, a crowd started to converge in one of the open areas. Twitter engineers were mapping out the technical details of the company's backend architecture on a whiteboard for Musk's outside engineers. Soon, the CEO walked up and joined the crowd. Unlike the code reviews, the whiteboard session was thorough and lively. It seemed like the Goons earnestly wanted to understand how Twitter worked.

Then, just before midnight, Lin's manager called him to the front of the room. "Randall, you need to explain to Elon how the heavy ranker works," he said, referring to part of the ranking algorithm. Lin picked up a marker and started mapping out a high-level overview. Musk seemed pleased. "Cool, how can I help?" Musk asked when Lin finished.

Lin had been thinking about what to say to Musk all evening. He seized the opportunity to pitch his personal obsession to his new boss. "First, we need to own our own GPUs," he said, referring to the prized computer chips that process data very fast. "We're training our machine-learning algorithm in the cloud. We need to own the machines if we are going to take the next step." "Done," Musk said. Then Lin suggested canceling the Google Cloud contract. "We're spending hundreds of millions of dollars a year and we don't need to be," he said. Musk looked ecstatic.

In many ways, it had been the perfect display for the new CEO: a highly technical show of ambition, paired with clear cost savings. Lin had pushed for owning the GPUs for years, and he'd finally found a receptive audience.

As he wrapped his presentation, Lin asked Musk who he should send the bill of materials to, in order to get the GPUs. Musk said simply: "Me."

The timing couldn't have been better. In the months that followed, the market for GPUs exploded as AI startups scrambled to get their hands on the valuable chips. One AI startup CEO compared the technology to a "rare earth metal." Shares of Nvidia, one of the main providers of GPUs, shot up 315 percent from October 2022 to April 2023 in response to skyrocketing demand. Lin was hardcore, and he was also ahead of the fray.

CHAPTER 37

"Vox Populi, Vox Dei"

As Lin was going all in on Twitter 2.0, Yoel Roth was getting ready to tell the world why he'd resigned. "Since the deal closed on Oct. 27, many of the changes made by Mr. Musk and his team have been sudden and alarming for employees and users alike, including rapid-fire layoffs and an ill-fated foray into reinventing Twitter's verification system," Roth wrote in a November 18 op-ed in *The New York Times.*

In the article, Roth explained the limits of Musk's vision for free speech absolutism. Regardless of who owned the company, Twitter was bound by regulators, advertisers, and, crucially, Apple and Google, the two tech giants that controlled the smartphone ecosystem. If the company failed to police harmful content, it could risk expulsion from the app stores—a death knell for any online platform.

"In the longer term, the moderating influences of advertisers, regulators and, most critically of all, app stores may be welcome for those of us hoping to avoid an escalation in the volume of dangerous speech online," he wrote. "Twitter will have to balance its new owner's goals against the practical realities of life on Apple's and Google's internet—no easy task for the employees who have chosen to remain. And as I departed the company, the calls from the app review teams had already begun."

Musk didn't want Twitter to get kicked off Apple's App Store. But he was willing to push the limits of what the company would accept. For months, he had been toying with reinstating banned accounts, including that of former President Donald Trump.

The move was more or less symbolic. After Trump was kicked off the platform, he'd helped start a right-wing Twitter alternative called Truth Social. The app had never taken off, but Trump's agreement with the company had an exclusivity clause, meaning he had to wait six hours before sharing content on a different platform.

The same day Roth published his op-ed, Musk asked his followers to weigh in on the decision. "Reinstate former President Trump," he wrote, alongside an interactive poll. More than 15 million accounts voted, with 51.8 percent of respondents saying yes.

"Vox populi, vox Dei," Musk responded, meaning, "the voice of the people is the voice of God." Trump's account was coming back.

The move presented a problem for Twitter engineers. It wasn't as simple as flipping a switch. Every time an account was reinstated, they had to rebuild a social graph, activating data on who the account followed and who followed the account. Every time someone followed, blocked, or muted another user, that connection was recorded. For an account like Trump's, with eighty-eight million followers, that was millions of pieces of data that Twitter had to update and maintain. Twitter didn't have the manpower to make sure the reinstatement went off without a hitch.

"Urgent escalation request—please DM me immediately," an engineering manager wrote in Slack moments after Trump's account was supposed to go live.

"What happened here?" asked Christopher Stanley, one of Musk's trusted lieutenants from SpaceX.

It turned out the reinstatement hadn't worked. The team tried again. Success. Then they realized the "follow" button on Trump's account wasn't working. They tested the issue with Musk's account. It turned out the issue was specific to Trump.

The engineering team scrambled to find someone, anyone, who knew how to fix the issue. Normally, they'd have engineers on call who'd know exactly what to do. But those people had all been laid off. The issue was escalated to director-level employees.

Ella Irwin, who'd taken over as the head of trust and safety after Roth left, popped into the Slack channel. "Hello everyone," she said. "Thank you so much for working on all this. From Elon: anything we need from Elon to help here? If not that is ok but wanted to confirm."

The engineers were still heads down trying to figure out what was wrong. "Trump follower count return [sic] different results from each refresh," one said, posting a side-by-side photo of Trump's account, each with a wildly different number of followers.

"It works for me now," an engineer responded. "I hit follow, I refresh, it says following."

"Same here," piped in Irwin.

Soon after, Trump announced that he remained committed to his own social media platform, Truth Social, and, despite Musk's efforts to reinstate him, saw no reason to return to Twitter.

Musk followed up with another interactive poll five days later. He asked his followers whether Twitter should "offer a general amnesty to suspended accounts, provided that they have not broken the law or engaged in egregious spam."

This time, the response was unequivocal. More than 70 percent of respondents said yes.

"The people have spoken," Musk said. "Amnesty begins next week. Vox populi, vox Dei."

JP Doherty was sitting down to eat dinner when he got a message on Slack saying the company was bringing back sixty-two thousand formerly suspended accounts. One of the accounts had more than five million followers; seventy-five had more than a million.

Doherty warned that Twitter needed to be ready to turn on an account with five million followers—an event he referred to as the "big bang." He didn't want the accounts reinstated, but no one was asking for his opinion, just his technical expertise.

With Doherty's reluctant help, Musk followed through on his promise, reinstating thousands of accounts, including those belonging to Andrew Anglin, founder of the neo-Nazi website Daily Stormer, and far-right extremist Patrick Casey, who was subpoenaed by Congress for his involvement in the Capitol riot.

The move alarmed Twitter's advertising team. In December, employees won a small victory when Musk agreed to stop the reinstated accounts from being monetized. Moving forward, Twitter would add a "silent reject" label to prevent them from promoting ads, at least until the advertising crisis had blown over. On Slack, an employee explained that this would not give the accounts "any auto-rejection messaging which we want to avoid." Ironically, it sounded like a shadowban.

CHAPTER 38

"Our Concerns Regarding the Quality of Your Coding Ability"

On November 21, employees received a new email from Musk. Every Friday, he wanted them to send an update on what they'd worked on that week and what they were trying to accomplish. For engineers, he also wanted to see a code sample.

Since the takeover Musk had made it abundantly clear that he didn't trust his new workforce. One week, he was pressing engineers to launch new features; the next, he was firing people "for cause" to keep costs down. This move seemed to fall into the latter category, aimed at weeding out more workers.

"The reason Twitter sought to engineer resignations or excuses for for-cause firings is clear: were Twitter required to keep its word to all of the laid-off employees and actually pay them severance per the pre-existing policy, the total cost would easily be in the nine figures," an employee lawsuit, Arnold v. X Corp., later alleged.

Musk had already warned managers not to protect low performers. "At risk of stating the obvious, any manager who falsely claims that someone reporting to them is doing excellent work or that a given role is essential, whether remote or not, will be exited from the company," he wrote in an internal email.

No sooner had engineers sent the code samples than they started receiving warnings. "The purpose of this written performance warning is to bring to your attention our concerns regarding the quality of your coding ability," the emails began. "And to define for you the seriousness of the situation so that you may take immediate corrective action." Failure to meet expectations could result in termination, the email read. "Please use this opportunity to restore confidence and demonstrate your contributions to the team and company."

Then, the evening before Thanksgiving, roughly fifty engineers were fired. "As a result of the recent code review exercise, it has been determined that your code is not satisfactory, and we regret to inform you that your employment with Twitter will be terminated effective immediately," the note read.

In exchange for signing separation agreements, engineers were offered four weeks of severance, which, they suspected, Musk would fight not to pay.

CHAPTER 39

"Exceptionally Poor Taste"

Twitter was losing money, fast. In late November 2022, weekly bookings (the number of brands that "booked" ad spots on Twitter) were down 49 percent week over week, according to a Slack message sent by a revenue financial analyst. In the last two months of the year, spending by the top thirty advertisers on the platform dropped precipitously to around $54 million, according to one estimate. Twitter's revenue in the fourth quarter of 2022 was a little over a billion dollars, down 35 percent from the year before and just 72 percent of the company's goal, according to an internal presentation from Twitter ad exec Chris Riedy in January. The presentation showed that the company was on track to meet its Q4 revenue goal until Musk took over. Considering that 90 percent of Twitter's revenue came from advertising, any shortfall could be devastating.

Twitter Blue was supposed to help fill the gap. But employees were growing increasingly concerned about the project's ability to make money. After pausing new sign-ups, Musk said the subscription service would relaunch on November 29. Then he said the project was on hold "until there is high confidence of stopping impersonation." The delay gave employees time to ponder the economics of the initiative—and what they were seeing didn't make sense.

Musk had promised Twitter Blue subscribers that they'd see fewer ads than regular users. "Since you're supporting Twitter in the battle against the bots, we're going to reward you with half the ads and make them twice as relevant," read the release notes (again, sloppily written by Jason Calacanis).

Internal estimates found that this plan would cost Twitter $6 per user per month in ad revenue. For subscribers who signed up via Apple's App Store, the company took a 30 percent cut of the subscription revenue, meaning each iOS subscriber cost the company 40 cents. As it stood, Musk's subscription service, priced in a spur-of-the-moment tweet at Stephen King, was going to *lose* the company money.

A group of hardcore employees, including Luke Simon, senior director of engineering, had a plan: require all Twitter users to opt in to personalized ads.

In 2021, Twitter's advertising revenue had taken a hit when Apple rolled out its App Tracking Transparency (ATT) feature. ATT required apps like Twitter and Facebook to ask users to opt in to app tracking, rather than having it turned on by default. If a user said yes, the company could track them across different apps and services, building out a more robust customer profile. But in practice, most people said no. After ATT went live, Meta estimated it would cost the company $10 billion in annual ad revenue. At Twitter, 34 percent of users opted in to app tracking.

But what if Twitter stopped giving users a choice? In an internal document about the project, the hardcore employees laid out a plan to force users to opt in to personalized ads. To keep using Twitter, users would need to give the company access to their location, allow it to share their data with business partners, and allow it to use "contact data from two-factor authentication (email, phone number) that we may have but not available for identity bridging, ad targeting, or prediction." Incidentally, using contact information supplied for two-factor authentication to target ads was the very thing the FTC had fined Twitter $150 million for in May 2022.

The document suggested that Twitter users could avoid sharing data if they signed up for Twitter Blue. This was bound to put the company in hot water with Apple, which prohibited apps from forcing users to pay to disable ad tracking.

I reported the details of the document in December, alerting Apple to Twitter's risky plan. Eventually, the project was abandoned. Behind the scenes, Musk was becoming increasingly incensed with how much power the tech giant had over his new company.

Apple has mostly stopped advertising on Twitter," Musk tweeted on November 28. "Do they hate free speech in America?"

The post inflamed the relationship between the world's richest man and the world's most valuable company. In a subsequent tweet, the CEO accused Apple of "threaten[ing] to withhold Twitter from its App Store" without explanation.

Like many of Musk's remarks, the accusation had a grain of truth to it, but the thrust of his argument was misleading. App Store guidelines suggested apps should not include "content that is offensive, insensitive, upsetting, intended to disgust, in exceptionally poor taste, or just plain creepy," and emphasized the need to protect children. In 2021, Apple booted Parler, a free speech alternative to Twitter, off the App Store for failing to adequately police hate speech and other harmful content.

Every few weeks, Twitter submitted an app update to Apple, which the iPhone maker had to approve before it could be released into the wild. After Twitter Blue launched, Apple reached out with pointed questions about impersonations. It hinted that if Twitter didn't crack down on posts that broke the rules, its next app release might be delayed, blocking the company from fixing bugs and rolling out new features. Ominously, on November 20, Phil Schiller, the Apple executive in charge of the App Store, quietly deactivated his Twitter account.

Historically, Apple might have brushed off Musk's remarks and refused to respond. The company was famously brand-conscious. The last thing it wanted was to get in a public brawl with Elon Musk.

But his accusations had hit a nerve. In recent years, Apple had come under antitrust scrutiny, specifically for the way it ran its app ecosystem. The tech giant was facing increased pressure from app makers who were angry about the 30 percent fee that it collected from in-app purchases. Epic Games, the company behind *Fortnite*, had sued Apple, arguing the fee was anticompetitive. The judge ruled in favor of Apple on most counts, but the company was still wary of renewed antitrust allegations.

Apple CEO Tim Cook, famous for his diplomacy, which had helped the company succeed in China, invited Musk to join him at Apple's headquarters in Cupertino so they could discuss the dispute.

On November 30, Musk tweeted a video taken on Apple's campus. The two CEOs chatted about supply chains and Musk decided not to press the Apple CEO about handing over the private data of Twitter Blue subscribers. Cook said that Apple would not stop advertising on Twitter, according to Musk's biographer, Walter Isaacson.

"Thanks @tim_cook for taking me around Apple's beautiful HQ," Musk wrote. "Good conversation. Among other things, we resolved the misunderstanding about Twitter potentially being removed from the App Store. Tim was clear that Apple never considered doing so." Cook's diplomacy had once again paid off, as Musk stopped griping about Apple's 30 percent app store fee, at least in public.

CHAPTER 40

"This Will Be Awesome"

Elon Musk's interactions with Twitter's engineering team had left both sides deeply dissatisfied. The CEO seemed to believe that many of Twitter's problems stemmed from technical incompetence. The engineers believed that Musk's understanding of mobile software development was outdated. Unfortunately for them, Musk held all the cards. While the engineers couldn't replace their CEO, he could most certainly try to replace them.

In November, Musk hired the famous hacker George Hotz, credited as the first person to ever jailbreak an iPhone, to fix Twitter's famously bad search functionality.

The thirty-three-year-old hacker had turned down a job offer from Musk in 2015 to work on Tesla's autonomous driving system, but the two men continued to admire each other. After Musk demanded that his employees be "extremely hardcore" or quit, Hotz tweeted, "This is the attitude that builds incredible things. Let all the people who don't desire greatness leave."

Hotz offered to join Twitter for a twelve-week internship for the cost of living in San Francisco. He had recently decided to take some time away from his advanced driver-assistance company, Comma.ai, writing, "It's no longer a race car, it's a boat. And steering a boat requires too much damn planning and patience." Musk responded to his offer positively. "Sure, let's talk," he said.

Two days later, Hotz announced he was working at Twitter.

Hotz liked simplicity. What he saw at Twitter was unnecessary complexity. The hacker believed that fifty people could build and maintain the platform pretty easily. It sounded hubristic, but hubris had served Hotz well so far. He'd hacked the iPhone when everyone said it couldn't be done, then took on autonomous vehicles to compete with tech giants like Uber and Apple when everyone said he was crazy.

Twitter engineers were not impressed. Hotz had good ideas but struggled to put them into action, according to Randall Lin. He could also be kind of annoying. Dave Beckett, a site reliability engineer, recalled seeing a Slack message from Hotz at 10 p.m., asking if there was dinner in the San Francisco office.

Hotz told Musk to stop shipping features and focus on refactoring the code base. It would take at least six months, but at the end of it, Twitter's entire system would be smaller and far more simple, resulting in ten times less time for feature development, in his assessment.

Musk had no intention of slowing down on product development. Hotz's internship ended after just four weeks, leaving employees scratching their heads wondering why he'd been invited to Twitter and what he had actually done. Beckett summarized: "Weeks later he left with landing, we think, one commit," referring to the process of contributing to Twitter's code base. But Musk wasn't done bringing in outsiders to scrutinize his company.

What is he talking about? JP Doherty thought after he read Musk's latest tweet. It was 12:45 p.m. Pacific time on December 2, and the CEO of Twitter had just announced that he was about to drop a bombshell report on what had *really* happened with the Hunter Biden laptop story. Musk tweeted, "This will be awesome," with a popcorn emoji, but Doherty had a sinking feeling in his stomach. There was no way this was going to end well.

The story dated back to 2020, when the *New York Post* published a potentially explosive front-page story about Joe Biden and his troubled son, Hunter Biden, weeks before the US presidential election. The Rupert Murdoch–owned paper claimed that Hunter, who had long struggled with addiction, introduced his father, then the vice president of the United States, to an executive at a Ukrainian energy firm less than a year before the VP started pressuring government officials in Ukraine to fire a prosecutor who was in the midst of investigating the company.

The allegations were based on a trove of data that had been mysteriously recovered from Hunter's laptop. In addition to revelations about Hunter's business dealings in Ukraine, the files included nude images and a raunchy, twelve-minute video that apparently showed the Democratic nominee's son smoking crack while engaged in a sex act.

Trump's advisers Steve Bannon and Rudy Giuliani expected the 2020 story to crush Biden and swing the election decisively in favor of President Trump. They'd spent a considerable amount of effort working to make this happen. The timing and placement of the report wasn't an accident—it was the result of careful planning on their part, according to a thorough investigation by *New York* magazine.

But as the report zipped its way through message boards and social media, a theory emerged among tech policy experts that threatened to overshadow its racy revelations: the leaks looked like the result of a Russian hacking campaign.

In 2016, when Hillary Clinton was running for president against Donald Trump, WikiLeaks released a cache of emails from top Democratic campaign officials revealing deep ties to Wall Street and the Clinton Foundation's questionable ethics. The Justice Department, under Special Counsel Robert Mueller, conducted an extensive investigation, ultimately revealing that Russia was behind the attack.

In the wake of the election, which Clinton narrowly lost, the social platforms admitted just how far Russia had gone to try to sway the vote in favor of Donald Trump. Among other things, Russian bots had tweeted

2.1 million times during the campaign. Their disinformation posts on Facebook reached 126 million people.

"The effect of such manipulations could be momentous in an election as close as the 2016 race, in which Clinton got nearly 2.9 million more votes than Trump, and Trump won the Electoral College only because some eighty thousand votes went his way in Wisconsin, Michigan, and Pennsylvania," wrote Jane Mayer in *The New Yorker.*

Four years later, when the Hunter Biden laptop story started to circulate, executives at Facebook and Twitter stepped into overdrive, removing links to the article to stop it from going viral, eager to get ahead of what could be another Russian hack.

Yoel Roth, then Twitter's head of trust and safety, advised the company to stop there. He suggested a "freedom of speech, not reach" approach—allow the story to stay on Twitter but stop it from being recommended or amplified. But on this front, Roth was overruled. Twitter's hacked materials policy prohibited the distribution of content "obtained through hacking that contains private information, may put people in physical harm or danger, or contains trade secrets." Under that policy, Twitter took down the *New York Post*'s tweets about the story and froze the accounts of users who shared the article, including the *Post*'s official account and accounts belonging to members of the Trump campaign. Twitter users couldn't even share the story in direct messages with one another—an unprecedented move in the company's history.

Over the next day, though, Twitter executives determined they didn't have enough evidence to confirm that the materials had actually been produced through hacking. A further investigation outside the company revealed that Hunter had dropped off his laptop with a Trump-loving repairman, then failed to pick it up. The repairman contacted the FBI, then Congress, then Trump's lawyer Rudy Giuliani. Giuliani took the story to the *Post*.

"We made a total mistake with the *New York Post* and we corrected that within 24 hours," Jack Dorsey said when he was dragged before Congress in March 2021 to explain Twitter's reasoning. He sported a shaved head and his signature scraggly beard, a suit jacket over his black T-shirt. "It was not to do with the content. It was to do with the hacked materials policy. We had an incorrect interpretation," he said.

Roth echoed this viewpoint in his own House testimony in 2023. "Twitter erred in this case because we wanted to avoid repeating the mistakes of 2016," he said.

Musk felt there had to be more to the story. It was simply too convenient that Twitter, a company overrun with liberals, had played such a pivotal role in suppressing the scandal.

Musk gave Matt Taibbi, a writer who positioned himself in contrast to the mainstream press, indirect access to Twitter's internal systems, including emails and confidential documents. Soon, the project grew to include other writers sympathetic to Musk's position, such as former *New York Times* columnist Bari Weiss as well as antivax advocate Alex Berenson (also a former *New York Times* reporter), among others.

Doherty wasn't sure where the project was headed when he saw Musk's tweet. Theoretically, as the de facto head of Twitter Command Center, he was supposed to get briefed about big events *before* they happened. Lately, however, he'd been finding out about Musk's plans the same way everyone else did: by reading the CEO's tweets. He hoped Twitter could handle the spike in traffic.

Finally, around 3:30 p.m. Pacific time, after a few delays where Musk said the team was "double-checking some facts," the CEO handed the reins to Matt Taibbi. The journalist started tweeting:

1. Thread: THE TWITTER FILES
2. What you're about to read is the first installment in a series, based upon thousands of internal documents obtained by sources at Twitter.

3. The "Twitter Files" tell an incredible story from inside one of the world's largest and most influential social media platforms. It is a Frankensteinian tale of a human-built mechanism grown out the [sic] control of its designer.

Over the next hour and a half, as Taibbi's thread unfolded, Doherty grew more and more disgusted. He agreed with Musk that Twitter, like all big tech platforms, could benefit from more transparency. But Taibbi's thread didn't feel like an honest accounting of the company's past decision-making but rather a dishonest narrative aimed at pushing Musk's political agenda.

Taibbi argued that the documents showed a close relationship between the US government and Twitter's leadership, with government officials routinely asking for specific tweets to be taken down. He acknowledged that both parties had access to these tools, and that Twitter had honored requests from Trump and Biden staffers in recent years. But he claimed the Biden campaign had *closer* ties with Twitter because Twitter employees were left-leaning.

Taibbi correctly said that Twitter had taken "extraordinary steps" to suppress the Hunter Biden laptop story, including locking the account of then White House spokeswoman Kayleigh McEnany, after she tweeted about it. "The decision was made at the highest levels of the company, but without the knowledge of CEO Jack Dorsey, with former head of legal, policy and trust Vijaya Gadde playing a key role," he wrote.

He included screenshots of internal messages from Gadde, Roth, and other high-ranking employees who were involved in the *New York Post* decision. Taibbi characterized the team as blithely censoring the *New York Post* story without proof, yet the screenshots showed the leaders first agonizing over what to do, and then agonizing over what they had done.

"I'm struggling to understand the policy basis for marking this as unsafe, and I think the best explainability argument for this externally would be that we're waiting to understand if this story is the result of hacked materials," wrote the head of US policy communications after the company decided to block the links. "We'll face hard questions on this if we don't have some kind of solid reasoning for marking the link unsafe."

In a separate screenshot, Roth explained that "the policy basis is hacked materials—though, as discussed, this is an emerging situation where the facts remain unclear. Given the SEVERE risks here and lessons of 2016, we're erring on the side of including a warning and preventing this content from being amplified."

As part of his tweet thread, Taibbi published the name of a low-level incident manager in Australia who'd responded to a message from a member of Twitter's public policy team asking why McEnany's account was locked. The employee's message was anodyne. But publishing her name all but ensured she would be harassed and likely receive death threats. "Per checking, the user was bounced by Site Integrity for violating our Hacked Materials policy," the message read.

What the fuck? Doherty thought. The incident manager had no decision-making power. She'd just checked the log to see why the account was suspended and reported back. Immediately after Taibbi named her, Musk's supporters piled on, calling her a censor and claiming she was responsible for suppressing the *New York Post* story.

Doherty sent her a message on Slack explaining the situation and telling her to stay safe. Then he pinged the corporate security team.

"We need somebody to go check in with her and make sure she's OK," he said. "She has no idea what's coming."

It was a Saturday morning in Australia when the incident manager, who we'll call Sophie, learned that her name had been exposed. Sophie was in her first trimester of pregnancy and dealing with some house maintenance issues. If there was ever a good time to be doxxed, this was not it. Within hours, the threats started to pour in. On Instagram, on Twitter, on Facebook. Most were insults, some overtly racist (Sophie is Asian). One said she was an "example of how leftists abused the system." Another simply called her a communist. Sophie deactivated all her social media accounts except her Instagram, which was already set to private.

At first, she didn't think too much about it. It was unsettling, but she hadn't been part of the decision to lock McEnany's account. Then the messages started to get more serious. She got a death threat. People were blaming her, using her as a scapegoat for the fiasco.

Sophie reached out to corporate security but didn't hear back immediately. Over the next twenty-four hours, people found out where her husband's family lived in the Philippines and started posting links to their location on Twitter. Sophie contacted the trust and safety team, but nothing happened. She escalated the situation to Ella Irwin. Finally, Twitter started blocking links to the site that contained her family's information.

On Monday, Sophie finally met with the corporate security team. Afterward, they shared a document advising her on how to stay safe. She also met with Irwin, who told her that it had been a mistake to leave her name unredacted.

Sophie took a one-month mental health leave from work. Even while she was out, she kept responding to emails about her and her family's security. "It felt like I had to take a lead on this," she told me, disappointed that Twitter hadn't taken the threats more seriously.

Soon after, Musk slashed Twitter's parental leave policy, from twenty weeks down to the minimum legal requirement plus two weeks. Sophie was about to be a mother; she'd expected to be with her newborn for four months. The policy change was cruel, and sudden. Her mental health started to get worse.

"I was crying every day, depressed and angry," she said.

She would leave Twitter two months later.

In the aftermath of the Twitter Files, two versions of reality emerged. In one, Musk was a hero, fearlessly sharing bombshell revelations that conservatives had been silenced under the old regime—by Roth and Gadde in particular.

In another, the Twitter Files were a nothingburger—"a desperate attempt to legitimize a well-worn conservative narrative that the suppression of Hunter Biden's 'laptop from hell' proved collusion between the so-called deep state and social media companies," as Joan Donovan, a research director at Harvard Kennedy School, wrote in *Politico*.

Doherty's views fell squarely in the second category. "It was dangerous," he said. "You have this group of people who already think there's a giant conspiracy against them and you're just fueling the fire. It was dangerous, and it was reckless. If you really want to do it, then throw up *all* the communications so people can see what really happened."

This view became more widely shared on the left over the next few months as Taibbi continued to publish his findings while misinterpreting some of the evidence. Take Twitter Files #17.

In 2021, one of the leading disinformation research labs in the United States, the Atlantic Council's Digital Forensic Research Lab (DFRLab), sent Twitter a list of forty thousand accounts as part of an investigation into coordinated disinformation in India. The investigation was being conducted with an independent Indian news outlet called *The Wire*.

The partnership was still in its infancy but appeared promising. Already, one of the paper's hotshot young journalists, Devesh Kumar, had compiled a list of accounts believed to be engaged in coordinated inauthentic behavior in support of the Bharatiya Janata Party, the ruling party under Prime Minister Narendra Modi. Kumar hinted that the BJP might be using social platforms like Twitter to manipulate public perception.

Except, when DFRLab sent the list to Twitter, the story didn't check out.

"I spot-checked a number of these accounts, and virtually all appear to be real people," Yoel Roth told the research group.

DFRLab dropped the story and terminated its partnership with *The Wire*. The following year, *The Wire* published its investigation, which claimed to show the BJP used a shadowy surveillance app to manipulate social platforms and hijack the public conversation.

Months later, the paper removed that story after a piece I reported in *Platformer* revealed that Kumar had fabricated evidence in a separate investigation into Meta—an odd convergence of events.

Taibbi included details about the list of suspicious names DFRLab had sent to Twitter in a Twitter Files thread about state-sponsored blacklists. In Taibbi's telling, however, the list was not proof that DFRLab had done its homework and walked away from a hoax. Just the opposite. Taibbi suggested that in sending the list to Twitter, the research lab was implying that the people named on the list were part of the BJP conspiracy, and DFRLab was doing this simply because some on the list were "real septuagenarian Trump supporters."

"Such listmakers are either employing extremely expansive definitions of hate speech, extremely inexact methods of identifying spam, or they're doing both in addition to a third thing: keeping up a busywork campaign for underemployed ex-anti-terror warriors, seemingly referring to DFRLab, "who don't mind racking up lists of 'foreign' disinformationists that just happen to also rope in domestic undesirables," Taibbi wrote.

In the days and weeks to follow, Musk would continue hyping the Twitter Files ("Tune in for Episode 2 of The Twitter Files tomorrow!," "Taibbi ftw"), but the mainstream press didn't pay much attention.

Nevertheless, in December, as Musk took the mantle of conservative folk hero from Trump, liberal pundits correctly predicted just how far the Twitter Files would go to fuel the right-wing narrative that conservatives were silenced on social media. "They're gonna keep bitching about it for years. It's going to be louder and emptier than the Benghazi hearings," wrote Dave Karpf, an associate professor at George Washington University. "The scandal is that their clever propaganda effort sank like a lead balloon. And that has to be SOMEONE ELSE'S fault. That's it. That's the whole thing."

The Twitter Files did not go unnoticed by the FTC. On December 13, 2022, regulators sent Twitter a letter demanding information—not about the platform's perceived biases, but about the journalists who'd gotten access to Twitter's sensitive company data.

CHAPTER 41

"Assassination Coordinates"

Dave Chappelle stood onstage in front of a packed audience at Chase Center in San Francisco when he announced he was bringing on a special guest. "Make some noise for the richest man in the world," Chappelle said, his voice booming. Elon Musk walked onstage, wearing a black I LOVE TWITTER T-shirt. A loud round of boos followed.

The two men knew each other socially (in 2021, Chappelle got Covid after hanging out with Musk and Grimes, prompting rumors that he'd gotten it from one of them). Over the past two years, Chappelle's jokes about trans people had come under repeated scrutiny for being insensitive, leading to heated discussions about hate speech and cancel culture.

The audience wasn't supposed to have access to their cell phones or smartwatches. As they'd filed through the doors starting at 5:30 p.m. Pacific time, event staff instructed them to put their phones in Yondr pouches, which would stay locked until the end of the show—not unusual for a set from the high-profile comedian. At least a few people didn't comply and filmed Musk's arrival, which saw him strutting out onstage and raising his arms awkwardly, as if expecting a warm welcome.

As the noise continued, Chappelle laughed. "Cheers and boos, I see," he said, continuing to pace. Musk followed him around the stage, a mic held limply in his hand.

"Weren't expecting this, were ya?" Musk said.

"It sounds like some of the people you fired are in the audience," Chappelle responded.

The boos extended for several minutes before Musk walked offstage. Shortly after, the video went viral on Twitter.

Once again, Musk had failed to read the room, perhaps not anticipating how far his popularity had fallen, even in a tech hub like San Francisco. Later, he tried to save face by claiming that the crowd had been "90% cheers & 10% boos (except during quiet periods).

"But, still, that's a lot of boos, which is a first for me in real life (frequent on Twitter)," he added. "It's almost as if I've offended SF's unhinged leftists . . . but nahhh."

Anyone recognize this person or car?" Musk asked on December 14. He tweeted out a video of a young man wearing a black facemask sitting in the driver's seat of a Hyundai. The driver of the car, Brandon Collado, was allegedly stalking Grimes, who had three children with Musk—X Æ A-12, Exa Dark Siderǽl, and Techno Mechanicus. The troubled man claimed that Grimes was sending him coded messages through her Instagram posts. The night before, he'd apparently located a car carrying X Æ A-12 in Los Angeles, and jumped onto the hood. Musk was rattled.

The following day, Musk banned people from sharing the real-time location of other users on Twitter. He implied that an account that shared the location of his private plane, @ElonJet, run by a college student named Jack Sweeney, was to blame for the security incident. The account had pissed Musk off for years. Once, he'd offered to pay Sweeney $5,000 to shut it down. More recently, Twitter had begun applying "visibility filtering" (what Musk previously decried as "shadowbanning") to the account so its tweets would not be recommended.

Los Angeles police told the *Washington Post* that there was no evidence to suggest Collado had used @ElonJet to try to locate his target. The location of Musk's jet was not the same as his real-time whereabouts. Beyond that, Collado seemed obsessed with Grimes, not Musk.

Musk had previously said his commitment to free speech was so devout that it even extended "to not banning the account following my plane, even though that is a direct personal safety risk." Now he changed his mind and did just that.

As journalists began digging into the story, Musk made a surprising move. He blocked links to the jet-tracking account on other social media platforms and banned the accounts of multiple high-profile reporters, including Matt Binder of *Mashable*, Donie O'Sullivan of CNN, Drew Harwell of *The Washington Post*, and Ryan Mac of *The New York Times*, who were actively reporting on the story.

On December 15, Matt Binder was at the *Mashable* office in New York scrolling Twitter when he saw that O'Sullivan had been banned. "I'm probably next," he thought. Binder had tweeted about the jet tracker, specifically showing that Musk was now using the same visibility filtering that Musk had criticized at Twitter 1.0 to suppress the @ElonJet account:

"Remember Elon Musk's first Twitter Files? the one about Twitter blocking links to NY Post's Hunter Biden story," Binder wrote. "Elon Musk is using the same thing to block links to @ElonJet on other platforms right now. The exact same thing (except old Twitter stopped doing it the very same day)." He posted the tweet over a side-by-side image showing the error message that popped up when people tried to share the Hunter Biden laptop story on October 14, 2020, and an identical error message that displayed when people tried to share a link to the @ElonJet account.

Binder tried to download his data but it was already too late. He'd been suspended. His account was in read-only mode, meaning he could scroll Twitter, but couldn't tweet.

Musk claimed that the accounts were banned for violating Twitter's new policy against location sharing. "Same doxxing rules apply to 'journalists' as to everyone else," he tweeted. "They posted my exact real-time location, basically assassination coordinates, in (obvious) direct violation of Twitter terms of service."

The evening of the suspensions, thousands of people gathered on a Twitter Space hosted by *BuzzFeed* reporter Katie Notopoulos. Suddenly, Musk's name popped up. "Elon, thank you for joining," Notopoulos said. "I'm just hoping that you can give a little more context about what has happened."

"Uh yeah, as I'm sure anyone who's been doxxed would agree, sharing real-time information about somebody's location is inappropriate," Musk responded. He added: "There's not going to be any distinction in the future between journalists, so-called journalists, and regular people, everyone's going to be treated the same. You're not special 'cause you're a journalist. You're a citizen."

Drew Harwell, one of the journalists who'd been banned, pushed back. While he'd been kicked off Twitter, a security loophole allowed him to join Twitter Spaces (Matt Binder had also been in the room but was mysteriously kicked out before Musk entered). "You're suggesting that we're sharing your address, which is not true," he said to Musk. "I never posted your address."

"You posted a link to the address," Musk argued.

"In the course of reporting about @ElonJet, we posted links to @ElonJet which are now not online, using the same link-blocking technique you've criticized about the Hunter Biden *New York Post* story in 2020. So what is different?" Harwell asked.

"It's no more acceptable for me . . . it's no more acceptable for you than it is for me," Musk stuttered. "It's the same thing."

"So it's unacceptable what you're doing?" Harwell asked.

"No," Musk said. "You dox, you get suspended, end of story, that's it."

Then Musk left the room. Moments later, the company temporarily shut down Twitter Spaces. Musk claimed the shutdown was needed to fix a legacy bug—presumably the one that allowed suspended users to join Spaces in the first place.

The episode alienated even some of Musk's most ardent supporters. "I think it was a bad decision, and I think that it represented . . . the least generous interpretation was that it represented deep hypocrisy," said David Friedberg on the *All-In* podcast. "Not just a few weeks ago did he say he would never delete that account [@ElonJet] but he also said he was buying Twitter to enable freedom of speech and freedom of expression and that he wouldn't come in and do the same sort of content moderation that was done by the old regime."

That same evening, Musk found a way to walk back the bans, asking his followers when he should "unsuspend accounts who doxxed my exact location in real-time." The "now" option received the most votes. The reporters could come back as long as they deleted the violative tweets. It was a victory—but a small one. Musk had successfully reframed the discussion to be about doxxing rather than his about-face on free speech.

Binder didn't receive an official explanation about why his account was suspended—and he wasn't asked to delete a tweet before his account was brought back. After he tweeted about this on December 17, Ella Irwin DM'd him: "Hi Matt, I saw your tweet regarding not knowing the reason for your suspension. I don't comment publicly on the reason for a user's suspension since that is up to each user to announce or not announce. . . . For reference, you had posted a tweet the evening of 12/14 which had an image with a link to an Instagram account containing live tracking information. That is no longer allowed with the recent policy update on 12/14 as this information can be used to identify an individual's location and poses a safety risk."

CHAPTER 42

"Epic Shit"

Randall Lin was not going to be distracted by the discourse surrounding his erratic boss. He was busy retraining Twitter's machine-learning algorithm on the new GPU servers and getting ready to move the machines to a data center in Sacramento. His days were long but fulfilling.

In November, Lin had received an exciting message from James Musk. "Hey, Randall, Elon suggested I connect with you to understand some of your ideas for how we can improve things at Twitter," the message read. "Do you have time to speak tonight or meet tomorrow?"

Lin had nothing but time to devote to Twitter. Every morning, he took a twenty-minute bus ride to the office, listening to music on his noise-canceling headphones. He stayed there until well past dinner, hitching a ride home with one of Musk's top engineering directors. Musk had told Lin that Twitter should be showing "epic shit" on the home timeline, and Lin was determined to deliver.

Lin liked the changes he was seeing at Twitter. The company was cutthroat and focused on execution. Old Twitter wasted time on the wrong things. "Two people told me that they couldn't close an interview loop until they'd interviewed at least three candidates, and one had to be an underrepresented minority," Lin said. "It sounded highly inefficient."

Lin agreed with Musk's view that employees should show up in the office. Every single day, if they could. Earlier that month, the CEO made the mandate crystal clear. He'd told engineers to work on the tenth floor of the office unless they received an explicit exemption.

And people still weren't showing up! It was like they were being asked to fly to Antarctica. *Ridiculous*, Lin thought. He knew that he had it easier than some of his colleagues, who had families and lived farther from the office. In fact, he'd helped two of his teammates write letters to try to get remote work exemptions. But this couldn't be the norm for everyone.

Musk claimed that the average occupancy in the San Francisco office was below 10 percent, and that Twitter was spending $400 on lunch for each employee. He'd shared the data publicly, on Twitter. Twitter's former VP of real estate and work transformation, Tracy Hawkins, pushed back on that claim, noting that she'd run the lunch program and the average cost was, at most, $25 per employee and attendance was between 20 and 50 percent in the offices.

To Lin, the specifics of the fight weren't important. "Clearly, this culture is better for technical projects," he said. To Lin, that was really all that mattered.

CHAPTER 43

"This Explains a Lot"

Twitter's Trust and Safety Council, a volunteer group of around one hundred human rights experts, was not happy. The council, which formed in 2016, advised Twitter on matters related to hate speech and child sexual abuse material (CSAM). Musk's decision to lay off thousands of employees, including content moderators, and reinstate accounts banned for hate speech seemed to fly in the face of everything it stood for.

In early December, research from the Center for Countering Digital Hate and the Anti-Defamation League suggested that, under Musk, "daily use of the n-word" was triple the 2022 average, while slurs against gay men and trans people were "up 58% and 62%, respectively."

The report was the final straw for three high-profile members of the council. On December 8, Anne Collier, founder and executive director of the Net Safety Collaborative; Eirliani Abdul Rahman, an online safety advocate and cofounder of Youth, Adult Survivors & Kin in Need; and Lesley Podesta, chair of the Young and Resilient Research Center at Western Sydney University, resigned.

"It is clear from research and evidence that, contrary to claims by Elon Musk, the safety and well-being of Twitter's users are on the decline," they wrote in an open letter published on Twitter.

Michael Cernovich, an alt-right media commentator, replied to Collier's tweet with a link to a *New York Post* article titled, "Twitter refused to remove child porn because it didn't 'violate policies.'" The article referenced a 2021 lawsuit in which a teenage sex trafficking victim sued Twitter for failing to take down child sexual abuse material.

"You all belong in jail," Cernovich said.

The reply echoed a baseless QAnon conspiracy that liberal elites are engaged in child sex trafficking. It wasn't logical (was Cernovich claiming that a group of child safety advocates was responsible for Twitter's alleged failings on CSAM?), but it fit squarely into Musk's views on the alleged incompetence of the company's former leadership.

"It is a crime that they refused to take action on child exploitation for years!" Musk said, replying to Cernovich's tweet.

The conversation devolved in a matter of minutes. Jack Dorsey jumped in to defend his record, calling Musk's comments "false."

"I wish this was false but my experience this year supports this," Ella Irwin, the new head of trust and safety, responded. "I fought hard to get funding to replace the people working on this who left in early 2022 and was told no. At one point there were 0 engineers and very few employees working on CSE and still no funding."

Irwin's comment surprised some of her former colleagues. "Quit lying Ella," replied Rumman Chowdhury, a former director of machine-learning ethics, transparency, and accountability. "I . . . OFFERED expert resources to you in both research and engineering. You chose to ignore me."

Another employee told *The Washington Post* that Irwin had joined during a hiring slowdown "and created more than 10 new roles, none of them in child protection."

Yoel Roth was similarly startled. Both he and Jay Sullivan, Twitter's former head of product, took child safety extremely seriously. In the spring of 2022, Twitter had considered monetizing adult content, to compete with platforms like OnlyFans, which was projecting $2.5 billion revenue for the year. Sullivan shut the project down. An internal document about the decision read: "Twitter cannot accurately detect child sexual exploitation and non-consensual nudity at scale."

Around that same time, Roth found a number of posts that included hashtags related to child sexual abuse material proliferating across the site. Some were spam. Others contained or linked to content containing nudity and were clear violations of Twitter policy. The company needed human content moderators to review every single post and report suspected CSAM to the National Center for Missing and Exploited Children, or NCMEC, per US law. If Twitter simply banned them and took down the posts, it would make it difficult for law enforcement to take action on the perpetrators.

Roth spent a weekend working with a senior engineer writing code to hide the hashtags across Twitter, making it difficult for people to find the posts while content moderators reviewed them.

Now, it seemed Musk and Irwin were implying that Twitter's former management had knowingly ignored the issue.

On December 10, a shadowy antitrafficking advocate named Eliza Bleu dredged up one of Yoel Roth's tweets from 2010 where he'd shared a *Salon* article talking about the age of consent. "Can high school students ever meaningfully consent to sex with their teachers?" Roth asked, quoting from the article.

"I think I may have found the problem @elonmusk," Bleu said.

"This explains a lot," Musk responded.

Then he shared a screenshot from Roth's thesis, where the former head of trust and safety, who is openly gay, called for tech companies to do more to protect minors on hookup apps like Grindr.

Except, that's not how Musk spun it. "Looks like Yoel is arguing in favor of children being able to access adult Internet services in his PhD thesis," Musk said.

Roth, who was sitting at his home in San Francisco, watched in horror as his phone exploded with death threats and homophobic slurs, not dissimilar from the torrent of filth the incident manager, Sophie, had received.

"Go fuck yourself you piece of shit," one man wrote, using what appeared to be his real name and email address. "You deserve to be hung you faggot."

"You're a traitor piece of shit hope you get prosecuted and sent to federal prison you're nothing more than a pencil neck dork piece of shit who betrayed his own country you're a the slimmest rat you and those other loser who worked at Twitter glad Elon took over and exposed your punk ass fuck geek lol," wrote another.

"fuck you roth, youre a lying, corrupt, mentally retarded, perverted, degenerate, child molesting pedophile bastard son of a dirty bitch whore cunt! you weren't birthed roth, youre bitch whore cunt mother had the shits and out youre bastard son of a whore ass squirted!" wrote yet another. It ended: "Sent from my iPhone."

There were hundreds just like it. Violent, homophobic, racist. The magnitude of hate was worse than anything Roth had ever experienced—and he'd been personally on the hook for the decision to ban Donald Trump.

The following day, Roth and his husband were packing to leave town for a few weeks when the *Daily Mail* published an article about their modest two-bedroom home in El Cerrito, which the paper called a "Bay Area mansion." The couple finished packing their bags, put their dog in the car, and fled. Later that month, they decided to sell their house, worried that if they returned, they'd be targeted. It was never clear how serious a death threat on the internet was. But why would you stick around to find out?

CHAPTER 44

"Deep Cuts"

As the due date of Twitter's first interest payment on the $13 billion loan inched closer, the Goons went on a rampage to cut $500 million from the budget. The company owed the banks $300 million in January. Between the interest payments and Twitter's other expenses, costs would total $6.5 billion in 2023 unless drastic action was taken.

Already the Goons had sent around a spreadsheet to a select group of managers and department heads detailing Twitter's expenses. It was so granular that it got down to the cost of employees' IVF treatments and personal security for a former executive. The Goons instructed managers to sign their names next to expenses they wanted to keep and include a note about why the expense was justifiable.

Steve Davis, CEO of The Boring Company, and Pablo Mendoza, managing director at Vy Capital, prepped the group on how to discuss the budget with Musk. They knew the areas he'd want to dig into, like office leases and software licenses, and wanted employees to be prepared.

On Saturday, December 10, the pair called a group of about forty department heads into the office to meet with Musk in person. "We were ready to show how hardcore we are," said one attendee.

The group spent the day going through the budget line by line. After one woman tried to argue for an item that Musk did not find necessary, he fired her on the spot. "You can be wrong, but don't be confidently wrong," he warned the group.

lumnist Bret Stephens became the main character after overreacting to being called a "bedbug" on Twitter. (He emailed the professor who wrote the tweet, challenging him to "come to my home, meet my wife and kids, talk to us for a few minutes, and then call me a 'bedbug' to my face." The professor, of course, promptly posted that email.) The sense of justice on Twitter can be imprecise. There's the woman who made a pot of chili for her neighbors ("presumptuous"); the wife who loves sitting in her garden drinking coffee with her husband every morning ("privileged"); the young adult author who rallied harassment toward rivals ("nepotism hire at defense contractor Lockheed Martin").

So why, then, did Elon Musk seem determined to be Twitter's main character every single day?

It's hard to know what exactly is going on in Musk's head at any time. But what is the one thing that the world's richest man can't seem to live without? Attention.

Throughout the day, employees learned which justifications appealed to Musk and which did not. The fact that Twitter had a contract in place was not a good enough reason to keep paying. Musk said the only place that made Tesla sign paperwork was the DMV and urged people to try to negotiate every deal down by at least 75 percent.

By the end of the month, the Goons succeeded in saving $1 billion in costs, double their original goal.

The infrastructure team, in charge of managing Twitter's supply chains and data centers, was doing its best to slash its budget. In early 2022, the team had planned to spend roughly $1.2 billion on new servers and associated energy costs. Now, Steve Davis told Nelson Abramson, global head of infrastructure, that he needed to get the budget as close as possible to zero.

Abramson relayed the information to his team. Some felt the request was preposterous. If pushed, they reasoned they could get the budget down by about half, to $650 million. But "as close to zero as possible" meant core engineering teams wouldn't get even a fraction of the new machines they'd been promised. Without new servers, Twitter could become glitchy and prone to bugs.

Some of Musk's Tesla engineers couldn't understand the problem. "All of Tesla runs on a single data center, why do you need three?" one asked a Twitter employee. Pointing out that Twitter was a global social network, and would require far more computing power and storage than a car company, seemed fruitless.

To appease the Goons, the team created a spreadsheet called "deep cuts" and populated it with made-up numbers, ballparking the new budget at around $125 million. They hoped that when engineering managers saw how few machines they'd be able to get they'd fight back—and the Goons would be forced to increase the budget.

"I was singing 'The First Cut Is the Deepest' and joking about what color to make the spreadsheet. That's how not concerned we were with the project, because it just seemed so ridiculous," one employee told me.

But when the Goons saw the deep cuts document, they didn't think it was a joke. In fact, they were thrilled. "I got word that the Goons *really* liked this project," the employee said. Suddenly, the budget that had started as a joke became reality.

CHAPTER 45

"We Just Won't Pay Those"

The Goons' budget-cutting initiative did not bode well for the real estate team, which was in charge of Twitter's global office leases, an annual expense of $130 million. Tracy Hawkins, vice president of real estate, was growing increasingly concerned.

Hawkins hadn't known much about Musk when he'd proposed buying the company back in April. Throughout the summer, as his behavior became more adversarial, she'd tried to keep an open mind, hoping that once they were all on the same team he'd change his tune.

Hawkins had more than twenty years of experience in commercial real estate. She was on the board of the Northern California chapter of the professional association CoreNet Global. She prided herself on having a sterling professional reputation. In real estate, reputation was everything. If landlords didn't trust her, how could she negotiate good leases for Twitter?

In October, Hawkins had met with Steve Davis and Jared Birchall and learned Twitter would no longer be working with brokers to negotiate its office leases.

The move went against best practices for the industry, but Davis and Birchall made it clear the decision wasn't up for debate. Hawkins was simply told, "Elon wants this."

Hawkins was given a spreadsheet of expenses and told to identify leases that could be canceled. She dutifully filled it out, adding context about lease termination fees that might be incurred if Twitter walked away from certain agreements.

Davis did not take this information well, Hawkins alleged in a later lawsuit, Arnold v. X Corp., which she filed with a group of five former employees, including Joseph Killian, Twitter's global head of construction and design. "Well, we just won't pay those," Davis said, according to the employees' lawsuit. "We just won't pay landlords."

Already, Twitter had fired much of the janitorial staff in San Francisco, and the offices were starting to smell. "Inside on the 10th floor it started to really stink as garbage, toilets were dirty," a site reliability engineer named Dave Beckett later tweeted.

Hawkins felt like she had no choice but to resign. "Her options were stark: resign, and leave with her professional integrity and reputation intact and without being involved in a crime, or remain and be complicit in and held professionally responsible for a scheme to defraud landlords out of rent or other fees admittedly due to them, destroying her own reputation and career in the process," her lawyers wrote.

After Hawkins left, her responsibilities fell to Joseph Killian, who now answered to the Goons and Nicole Hollander, who was Steve Davis's girlfriend. As far as employees could tell, Hollander didn't work at Twitter, but she was involved in discussions about office design.

Since October, Davis and Hollander had been living at Twitter HQ, along with their newborn child, whose due date had been October 19, according to a public baby registry. Lin was astounded when he saw Hollander holding the kid in the office. "That baby looks too young," he remembers thinking. "It literally looked so small, like it popped out yesterday."

On December 9, Pablo Mendoza informed Killian that Twitter would no longer be paying rent at any of its offices globally, according to the employee lawsuit. Killian tried to point out that if Twitter defaulted on its rent payments it would have no chance of negotiating its leases. "In response, Mendoza told Killian that Musk had decided that Twitter would only pay rent over [his] dead body," the lawsuit alleged.

Steve Davis complained that Twitter had a fraction of the employees of SpaceX but paid five times the annual rent. (No one mentioned that Twitter had employed more than twice as many people when it acquired its various offices.)

The employee lawsuit, Arnold v. X Corp., also stated that Alex Spiro, Musk's lawyer, "loudly opined" that it was unreasonable for Twitter's landlords in San Francisco to expect to be paid rent, given that the city was such a "shit hole."

Yet even as the Goons pressed Killian to stop paying rent, Musk was working to make Twitter HQ a more comfortable place to spend the night. Davis told Killian that Musk wanted to add a bathroom next to his office so that he wouldn't have to wake up his security team to walk him to the toilet, the lawsuit alleged. Killian explained that it would take time for Twitter to obtain the necessary permits. But Davis pushed back, saying, in essence, "We don't have to follow those rules," according to the lawsuit. Killian pointed out that it would be difficult to hire a licensed plumber if it didn't get the right permits. But Davis didn't seem to care. He told Killian to hire an unlicensed plumber. Thoroughly bewildered, Killian explained that using an unlicensed plumber was a violation of the company's lease.

"Davis responded that management did not care about any of this, that they weren't interested in ensuring that the work was performed in accordance with the standards required by the lease, by the City of San Francisco, by the State of California, or any other authority, they just wanted it done," the employees' lawsuit alleged.

Musk also wanted to turn a series of conference rooms on the ninth floor into hotel rooms for employees. Killian claimed in the lawsuit that he was told "the 'hotel rooms,' soon renamed to 'sleeping rooms,' to avoid triggering the suspicions of the city inspectors, were just being installed to give exhausted and overworked employees a place to nap."

Davis also told Killian to start planning for en suite bathrooms. In response, Killian emailed the transition team and explained that so far the changes he'd made were just related to furniture, but the new changes would require additional permits.

Nicole Hollander paid Killian a visit and "emphatically instructed him never to put anything about the project in writing again," according to the lawsuit. "Hollander appeared surprised and distressed that Killian did not inherently understand that this was not a project for which Musk and the Transition Team wanted a written record."

Someone filed a complaint with the city. On December 7, city inspectors came by the office to investigate. "This is just furniture!" they said when they saw the rooms, according to the lawsuit. "We expected more drastic changes."

Killian stayed quiet about the bathrooms.

Tired employees were now regularly sleeping at the office. Killian was told to change the motion-sensor lights in the hotel rooms because they were bothering people at night. He submitted the request to the landlord, who said no.

Hollander told Killian to disconnect the lights himself, but he didn't feel qualified to do electrical work, according to Arnold v. X Corp. When he told Hollander, she berated him until he took her into the room and showed her the ceiling to explain why it simply wasn't safe. "Caught between a rock and a hard place, Killian hired an electrician to disconnect these rooms independently, putting Twitter in violation of both the building code and their lease." The requests continued piling up. Killian was told to put space heaters in the rooms—a fire hazard and a violation of Twitter's lease. Then Killian was instructed to put locks on the hotel room doors. California code requires companies to install locks that automatically disengage when the fire suppression system is engaged. But the Goons thought these were too expensive.

Killian resigned on December 10, 2022. The lawsuit attested: "Between the demands that he effectively participate in theft and fraud and instructions to take actions in violation of California law and that could put his colleagues' lives at risk in the event of a fire—a possibility only increased by the unlicensed use of space heaters—Killian had no choice but to walk away from the job he had dedicated over a decade of his life to."

CHAPTER 46

"An Absolute Scam"

As the Goons were busy slashing Twitter's costs, Musk was eyeing contracts between the social media giant and mobile carriers in countries like Indonesia, Russia, and India. These contracts, which involved the carriers supporting two-factor authentication, a security feature that asks users for an SMS security code at sign-in, were rife with fraud. Scammers made backdoor agreements with the carriers and created scores of bot accounts, in exchange for a cut of the carrier fees. The result was that Twitter was losing money *and* having to contend with useless bots. "This is an absolute scam, this is absolute BS," Musk said.

In some countries, Twitter was paying more to cellular carriers to support two-factor authentication than it was making in ad revenue, making operational costs a net loss, according to two employees.

On December 11, Musk tweeted a cryptic warning. "The bots are in for a surprise tomorrow," he said.

He directed engineers to block traffic from more than thirty major mobile carriers, including some of the largest telecoms in those countries. At first, Twitter targeted the subset of people who used two-factor authentication. Then it dramatically escalated the situation and blocked all traffic from the carriers.

"So that will block almost all of Indonesia and Russia, 64% of India, 60% of Malaysia, 50% of Laos, Iran and Iraq . . ." an engineer said on Slack after looking at the list of carriers.

The team followed orders. Moments later, complaints flooded in as users across Russia and India were suddenly unable to access Twitter.

A representative from Reliance Jio, the largest telecom company in India, reached out to ask what was going on. A Twitter engineer shared the note on Slack.

"I expect more emails like this to hit our peering queue tomorrow," the engineer said.

"We blocked a fair few huge carriers, so I would expect so," a colleague responded. They warned that Twitter would need to do business with Reliance Jio in the future. Disrupting the relationship could put Twitter's business in India at risk.

An hour later, Musk relented and Twitter unblocked the carriers, telling representatives from the companies that the issue was due to "routing configuration changes."

By now, the engineering team was used to Musk's impulsive management style. His priorities seemed dictated by what he saw on his feed. In December, Musk noticed a verified account using his avatar photo was promoting a cryptocurrency scam on Twitter. "He wants to know why it was missed," Ella Irwin wrote on Slack.

An employee explained that the account promoting the scam had been verified under Twitter's legacy verification system. The account had been hacked and the new owner was promoting crypto. Twitter had locked the account and removed Musk's photo from its profile.

The scam hadn't been immediately spotted because the tool Twitter used to detect suspicious activity, Smyte, was unstable. It was going down once a day and lagged after it was brought back up. Employees needed time to fix the issue.

Musk refused to acknowledge the fundamental tension between safety and speed that exists for any online platform. It would take thoughtful policies and proven technological solutions to protect users from the worst of what the internet had to offer. Ruthless efficiency couldn't solve every problem, especially when it came to people's safety.

But at Musk's Twitter, speed was paramount.

CHAPTER 47

"#TwitterDown"

JP Doherty had been working nineteen-hour days since November trying to keep Twitter stable, despite having a skeletal engineering staff. Twitter Command Center needed to have people on call twenty-four hours a day. Doherty wished he could find some semblance of balance—a day off with his kids, a second to not stare at his computer screen—but not if it meant making his colleagues work more than they already were.

In years past, many tweeps had taken time off between Christmas and New Year's, sometimes extending the vacation for fourteen days. Now, that type of paid time off wasn't an option. "Hey team, many of you have been asking me or your [engineering managers] about PTO policy in December," an engineering manager wrote on Slack. "While we are waiting for the new PTO policy from HR, Twitter management has decided that two weeks PTO leave is not acceptable. It is expected that most of us will work through December but with some good downtime around Xmas and New Years."

Doherty had already been planning to take the Christmas shift for TCC. His kids didn't care much about the holiday and he was happy to give his colleagues the time off.

Then, on December 23, Doherty got a call from Musk's lieutenant Sheen Austin, who told him to run a test to shut down one of Twitter's three data centers. At the time, Twitter was spending $115 million on power and space at the facility, which was located in Sacramento. Twitter leased the space from a company called NTT.

The test itself wasn't hugely concerning. Doherty and his team had spent years working to make sure Twitter could remain stable if one of the data centers went down. Each week, Doherty did a failover test, shifting traffic from one data center to another. If one of them started to buckle, he'd tweak Twitter's systems to make sure the company had enough capacity.

The issue was more than theoretical. In September 2022, when temperatures in Sacramento hit 115 degrees, Twitter's data center overheated, and Doherty's team had been forced to shut it down. Luckily, with three data centers, each individual site only needed to handle 120 percent of its normal load (each data center needed to handle 60 percent of the daily peak load of another center). "If we lose one of those remaining datacenters, we may not be able to serve traffic to all Twitter's users," Carrie Fernandez, then a VP of engineering at the company, told employees.

Now, Doherty worried that if the test was successful, Musk would shut down the site permanently, forcing the remaining two data centers to handle 200 percent of their normal traffic load, and making Twitter less stable in the process.

Twitter had significant technical debt from its early days. The quirks of how the code was originally written meant that some of Twitter's services were hardcoded into specific machines in Sacramento. Because of that, testing the shutdown was crucial. Doherty couldn't just shift everything over to the other data centers automatically. He needed to make sure everything was running smoothly so there wasn't a disruption in Twitter's service.

As if on cue, Austin called back later that night. It wasn't a test—Musk wanted the site shut down permanently. Doherty felt sick to his stomach. The project would require his whole team, many of whom planned to take the holiday off.

"Dude, no part of this ends well. You're squandering any goodwill you have at this company if you do this," Doherty told Austin. His team had been working around the clock. They needed to rest. They'd earned it. But Austin wouldn't budge. Doherty reluctantly called his colleagues.

Randall Lin was working at the office late on December 23 when he ran into James Musk and his brother, Andrew, both of whom now worked at Twitter. They told him they were about to head out for the holidays. They were going to Austin, Texas, to spend time with Elon's family. Lin said he was about to leave for the airport, too.

A few hours later, Lin got a message from James Musk and his friend Ross Nordeen, who also now worked at Twitter. They asked if he knew where the Sacramento data center was. For security reasons, the exact location of Twitter's data center was a tightly held secret. Lin gave them the contact information of someone who knew.

The Musks didn't get the holiday they'd planned. That night, they were flying with Elon on his private jet when the topic of shutting down the Sacramento data center came up. Musk wanted to get out of the contract with NTT, but a Twitter employee had told him it would take six to nine months to move the servers. "Why don't we just do it now?" James Musk asked, according to Walter Isaacson. The jet made a U-turn in the sky. They were heading straight for Sacramento.

The group arrived late that night and rented a Toyota Corolla, the only car that was available on such late notice. They drove to the data center. Inside, there were fifty-two hundred refrigerator-size racks, each with about thirty servers. "These aren't going to be hard to move," Musk said. An employee who worked at the data center pointed out that they'd need a contractor to lift up the floor panels to get the racks out the door. Musk turned to one of his security guards. "Give me your pocketknife," he said. He got to work.

Between Christmas and New Year's, Elon, James, and Andrew, along with a few other employees they'd recruited to work over the holidays, moved seven hundred racks out of the data center in Sacramento.

A Tesla employee who was working at Twitter went to the Apple Store in San Francisco and bought the entire stock of AirTags so they could track each rack in transit. They hired Extra Care Movers to haul the servers to Twitter's Portland data center.

On December 24, the CEO proclaimed proudly that Twitter was still working "Even after I disconnected one of the more sensitive server racks." The project looked like a success on his terms.

Four days later, Twitter experienced a major outage. #TwitterDown started trending on the platform.

Part III

MAIN CHARACTER

CHAPTER 48

"One Main Character"

The user @maplecocaine once tweeted: "Each day on twitter there is one main character. The goal is to never be it."

It is, arguably, the tweet that best captures Twitter's dynamics. Like many social platforms, Twitter signals engagement through a number of public-facing metrics. For Twitter, those are "Likes"—if someone approves of a tweet, they can signal this by clicking a heart; a user can go a step further and "retweet," republishing to their own account and immediately to whoever follows them. Every tweet shows the number of Likes and retweets, making it easy and obvious to see how much interaction a post has received.

Naturally, Twitter incentivizes users with these metrics. This isn't unique to Twitter—virality exists on competing networks like Facebook, Reddit, and even Pinterest. But because of the velocity of Twitter's engagement mechanics, the Notifications tab can really pop off, and, since most accounts are public, even small creators can reach an audience many multiples of their follower count. Since @maplecocaine posted that tweet in 2019, it has been retweeted over thirty thousand times, and earned its own dedicated entry on Know Your Meme, cementing the tweet as internet canon.

Things that tend to take off on Twitter: breaking news, self-righteous political opinions, outrage, and funny jokes, especially humor that is strange or edgy. Unfortunately, what constitutes "funny" is often a matter of opinion, and a miscalculated off-color joke can quickly land a user in hot water. One infamous example is a 2013 tweet from Justine Sacco, then a senior director of corporate communications at the prestigious holding firm IAC: "Going to Africa. Hope I don't get AIDS. Just kidding. I'm white!"

Read one way, it's a self-effacing joke about white privilege; read another, it's racist. Sacco, who had only 170 followers at the time, stepped on a plane almost immediately after sending the tweet. While she was airborne for eleven hours (she was, indeed, headed to South Africa), the tweet went viral. Users started blasting Sacco for her insensitivity and called for her employer to fire her. People piggybacked off it to call out the dangers of structural racism, or simply to make their own jokes. #HasJustineLandedYet started trending on Twitter.

The day after she touched down in Cape Town, the optics seemed to have gotten too bad for IAC. She was fired.

For better or worse, this is a key part of Twitter's allure. It encourages its users to go viral, rewarding them with an algorithmically baited audience. Twitter, in a sense, can turn a nobody into someone very popular overnight. It can also turn anyone into the day's public enemy.

"The platonic ideal of a Main Character tweet: A reasonable, low-stakes gripe about a relatable situation," podcaster Michael Hobbes wrote, "and 10,000 people in the replies making up scenarios so they can call her a terrible person."

Justine Sacco's tweet wasn't exactly a low-stakes gripe. But on December 20, 2013, she became Twitter's main character. She wasn't the first, and certainly not the last.

Since then there have been many main characters. There's "Bean Dad," the musician/podcaster who posted a self-aggrandizing thread about how he found a "teaching moment" for his daughter when she didn't know how a can opener worked. (Instead of showing her, she spent six hours trying to figure it out—users found this cruel, flipping the narrative to argue that this behavior was abusive.) Conservative *New York Times* co-

CHAPTER 49

"VIP Users"

In January 2023, Elon Musk's engagement on Twitter was slumping. James Musk asked a data scientist to investigate Elon's theory: Had a fired member of the engineering team hacked the algorithm to purposefully suppress his account?

The answer was simple: no. But the issue wasn't all in Musk's head. His tweets really weren't performing as well as they used to. The CEO had 124 million followers on Twitter, more than any other user on the platform. But according to a *Washington Post* analysis, the typical daily view count on his posts had dropped from 231 million in December to 137 million in the first six weeks of the year.

The CEO's mentions were full of right-wing influencers voicing similar concerns. Talk-show host Dave Rubin and the pseudonymous shitposter @catturd2 complained their engagement had tanked since the takeover. James Musk asked the data scientist to investigate those accounts, too.

The data scientist looked at the metrics, half hoping he would find proof of wrongdoing. Elon clearly wanted to blame the drop in engagement on sabotage. But he couldn't find any evidence to back it up. Instead, the numbers showed that Elon's account had gotten a significant bump in April when the deal was announced, then again in October when it went through. As the months went on, interest in the CEO had declined, as had interest in the accounts he interacted with. In other words, the drop in engagement wasn't rigged. It was largely organic.

Reluctantly, the data scientist told James what he'd found. James didn't want to hear it. "I get that the data says this, but Elon thinks something is wrong," he said, according to the employee's recollection. "I trust his intuition more than the data."

Randall Lin decided to take matters into his own hands. James was right—Elon's intuition was usually spot-on. The CEO could be shaky on the details, but the thrust of his argument was usually on the money.

Lin created a list of users, including Elon, venture capitalist Marc Andreessen, and *The Daily Wire* founder Ben Shapiro. Then he added in some left-leaning people to balance it out: President Joe Biden, Representative Alexandria Ocasio-Cortez, LeBron James. Finally, he threw in his own account, and the account of the friend who'd tweeted about Musk's first day in the office. Like Elon, he and his friend understood Twitter intuitively.

At first, Lin named the list "canaries," suggesting that Twitter was a coal mine. His plan was simple. Track the accounts, see how they performed, and tweak the algorithm accordingly. As the list started circulating around Twitter, it got renamed to "VIP users."

Lin didn't realize it at the time, but the name change would have big implications for how Twitter treated this subset of high-profile users. What started as a way for Lin to track how changes to Twitter's algorithm impacted VIP users ended up creating VIPs, as engineers allowed the accounts to bypass controls that might otherwise have stopped them from showing up too often in people's timelines.

CHAPTER 50

“What the Fuck Is Going on with the App”

On February 7, 2023, a group of engineers shuffled into Musk’s office on the tenth floor of Twitter’s headquarters in San Francisco. Musk sat scrolling on his phone. He didn’t look up when they entered. “What the fuck is going on with the app, guys,” he said flatly. It was clear he was talking about his engagement.

Yang, a soft-spoken programmer who’d been at Twitter for close to a decade, had been looking into the issue, much like Lin and the data scientist. He’d detected five possible reasons behind the drop, but the number-one reason seemed to be organic: people just weren’t as interested in Musk as they used to be.

Yang told Musk what he’d found. Despite the tense atmosphere in the room, he had the hint of a smile on his face. He was talking about data, his specialty. He called the issue a “popularity drop” and pulled up the Google Trends graph, showing Musk the jagged downward slope that mirrored his decline on the platform.

Musk’s hands were starting to shake. Yang didn’t notice, but Lin did—he was watching Musk closely and saw the crash coming from a mile away. *Shut up, man*, he wanted to yell. *Just stop talking.*

When Yang finally did stop talking, Musk fired him on the spot.

The next day, Lin walked into a meeting with Musk and the other engineers who still had their jobs. This time, he was determined to give Musk an answer. He'd studied the VIP list, "the canaries," and had a theory.

Twitter's recommendation algorithm had always been a black box even to those who worked on it. At a basic level, it worked like this: when users opened the app, the algorithm looked at a multitude of signals to determine what they should see.

The first step of the process was called "candidate generation"—essentially choosing which tweets to surface in the limited space on someone's screen. Typically, in the For You tab, those candidates would be drawn from accounts the user followed, accounts outside the user's network, and ads. Then the recommendation system applied a number of filters to ensure that a user's feed didn't include too many tweets from one account, or tweets that violated the company's community guidelines, or were from an account that person had blocked.

During this process, Twitter was scoring thousands of possible tweet candidates into a list of about thirty that would show up in the user's feed. Scores dictated whether or not a tweet was displayed and where it appeared on the list.

Lin explained that the ranker—the part of the process where Twitter scored the tweets—had always been poorly understood. But it was sensitive to negative engagement, meaning it was highly responsive to blocks and unfollows. The theory was that because Musk's account got a lot of negative engagement, it could be impacting performance. Lin was going to tweak the algorithm to see.

After Lin was done talking, Musk appeared mollified.

"Yeah, I told you it was something like that," he said.

CHAPTER 51

"Engagement Night"

Four days later, Elon Musk sat in a sky box at State Farm Stadium watching the Super Bowl in Glendale, Arizona, with News Corp CEO Rupert Murdoch and his daughter Elisabeth Murdoch. It was a balmy day, 75 degrees, and the Kansas City Chiefs were up against the Philadelphia Eagles. As usual, Musk split his attention, alternating between scrolling on his phone and watching the game.

Shortly after kickoff, Musk tweeted, "Go Eagles!!!" along with six American flag emoji. Less than an hour later, President Biden posted his support for the Eagles, too. "As your president, I'm not picking favorites," he said. "But as Jill Biden's husband, fly Eagles, fly." The text appeared above a video that showed Jill Biden wearing an Eagles jersey that read "Biden" over the number 46.

Few would dispute the fact that Musk was generally better at using Twitter than Joe Biden. But Biden's Super Bowl tweet was the clear winner. It was jokey and cute, an unselfconscious ode to his wife. Musk checked the view counts. His tweet had 9.1 million impressions. Biden's tweet had 29 million.

Then, around 8:15 p.m. Pacific time, the game ended. The Eagles had lost on a last-second field goal, 38–35. Musk was furious. He deleted his tweet.

Musk got on his jet and flew from Arizona to Oakland. He was headed straight for Twitter's office.

Monday, February 13, 2023, was a rare cloudy day in Santa Barbara. While Musk was traveling back from the Super Bowl, I'd been flying from Oakland to Santa Barbara, after a whirlwind reporting trip on which I'd somehow thought I could bring my one-year-old daughter and still get work done. I was glad to be home.

I was surprised, when I opened Twitter, to see an entire feed of Elon Musk. There he was, tweeting about Dogecoin. There he was again, commenting on a video of a shirtless man in the snow with two pickaxes. And again, talking shit about the press: "Vanity Unfair has fallen so far (sigh)." OK, what was going on?

"Is everyone else's entire For You page Elon replies 😅," I asked on Twitter. More than nine hundred people responded, most of them with variations of yes.

My phone buzzed. An unknown number was messaging me on Signal.

"Hi is this Zoe?" the message read. "I am a current Twitter employee and I want to share some details if you're interested."

In a past life, the message would've made me ecstatic. But I'd grown wary since reporting on Elon Musk. Nearly every time I published a story about the CEO, my messages flooded with angry dispatches from his fans. It wasn't hard to imagine that one might try to trick me by trying to pose as a possible source.

I asked for identity verification. When the employee sent over a badge and an ID, I asked what they wanted to talk about.

"Basically, over the past week, Elon has grown more frustrated with the engagement counts dropping; last week he fired an engineer over this," the employee said, referring to Yang. "He's been pushing all engineers to do investigations daily. Meetings are scheduled at 11 p.m. and often last till midnight. The issue we are solving is simple: why are Elon's tweet counts dropping. It's that and only that—not about other accounts."

I reached out to a handful of current employees to see what they knew. The more people I spoke to, the more insane the story got, as employees revealed the full details of what had happened the previous evening on "engagement night," which is what they called the post–Super Bowl work marathon that resulted in Twitter artificially boosting Musk's tweets.

I received a document titled "all hands on deck" in which the stated goal was to "figure out why engagement is different between these tweets" (the tweets in the document were Musk's and Biden's Eagles tweets, which were posted the day of the Super Bowl), along with a snapshot of Twitter's code that showed Musk's tweets were being boosted. More documents followed.

Once I had everything I needed, I told Casey Newton it was time to go. We hit publish.

YES, ELON MUSK CREATED A SPECIAL SYSTEM FOR SHOWING YOU ALL HIS TWEETS FIRST, the *Platformer* headline read. AFTER HIS SUPER BOWL TWEET DID WORSE NUMBERS THAN PRESIDENT BIDEN'S, TWITTER'S CEO ORDERED MAJOR CHANGES TO THE ALGORITHM.

The story was an immediate hit on Twitter, confirming what many had suspected but no one had been able to prove: Musk was rigging the game to favor his own account.

The next few days went by without any major news. Musk had more or less admitted that there was something fishy going on with the algorithm, tweeting the meme of force-feeding his tweets to regular users, and writing, "Please stay tuned while we make adjustments to the uh . . . 'algorithm.'"

But on February 17, the CEO came out swinging, publicly disputing my reporting for the first time. "The 'source' of the bogus *Platformer* article is a disgruntled employee who had been on paid time off for months, had already accepted a job at Google and felt the need to poison the well on the way out," he said. "Twitter will be taking legal action against him."

What the hell? I thought. *What is he talking about?*

All my sources for the story were current Twitter employees. Had the new source who'd reached out to me about the story left and gone to Google in the three days since the article had come out? I tried calling, but the source didn't answer. I called again. No answer. Now I was starting to panic.

Already, a reporter at *Insider* was reaching out, asking me and Newton if we wanted to comment on Musk's allegations.

Then I got an unexpected call from a contact at Whistleblower Aid, an organization that had coordinated two exposés with *The Washington Post*. They were concerned Musk was going to sue one of their clients, who was in fact a former Twitter employee who'd gone to work at Google.

On January 24, *The Washington Post* had published an explosive report detailing alleged privacy violations from a new Twitter whistleblower. The whistleblower claimed that Twitter's security controls were so relaxed that any engineer could access a feature called "GodMode," allowing them to tweet from any account—including the accounts belonging to Elon Musk and Barack Obama.

"In the past, there was a way to take the tweet service, run it yourself, and tell it to make a tweet as anybody," a former employee told me. "It would then send the tweet to the backend, and the backend didn't have any way of knowing that it wasn't the main tweet service." In other words, Twitter didn't have strong protections against internal attacks.

(Current Twitter employees were skeptical of the whistleblower's claims. The program he was referring to, which had since been renamed "preferred mode," was now carefully tracked. "Engineers are not incentivized to mess with user data any more than your UPS driver is motivated to contaminate your parcels," one told me. I had found this argument slightly ahistorical. In December 2022, a former Twitter employee had been sentenced to three years in prison for spying on behalf of Saudi Arabia. Clearly, the company *did* need to worry about internal threats, even if they were rare.)

Now, three weeks after the *Washington Post* article came out, I was seriously confused. Musk seemed to be implying that the GodMode whistleblower, whom I'd never spoken to and who was no longer working at Twitter, was somehow feeding me accurate information about what was currently going on at the company.

As I tried to work out what was going on, my source called back, and we were able to speak on FaceTime. They were still working at Twitter. They felt understandably nervous about Musk's tweet, but they had no idea what he was talking about.

Newton and I put out a forceful statement saying we stood by our reporting, and Musk never followed through on his threat.

In some ways, my story had already played right into his hands, making him the main character of the day. First, Twitter had been full of Musk's tweets; then it was full of tweets complaining that there were too many Musk tweets; and then my reporting explaining what had happened just drew attention back to him.

By the end of the week, did anyone even remember there had been a Super Bowl?

Zoë Schiffer @ZoeSchiffer—Feb 13, 2023
Is everyone else's entire For You page Elon replies 😅

Nicholas Brown @News_By_Nick—Feb 13, 2023
@ZoeSchiffer Yes

logan bartlett @loganbartlett—Feb 13, 2023
@ZoeSchiffer yes

David Weissman ✡️ @davidmweissman—Feb 13, 2023
@ZoeSchiffer Yes, WTF?

Santiago Pombo @SantiagoPombo—Feb 13, 2023
@ZoeSchiffer 100% but I was scared to bring it up.
I guess he does own the platform now 😬

Khaver خاور@thekarachikid—Feb 13, 2023
@ZoeSchiffer blocked him so no thankfully

Andrea Kuszewski 🧠 @AndreaKuszewski—Feb 13, 2023
@ZoeSchiffer This is how Elon 'fixed' the algorithm and remedied his low engagement numbers. Force himself down our throats I guess

anildash.com @anildash—Feb 13, 2023
@ZoeSchiffer My man is paying $10M per promoted tweet.

Brian Ray @brianrayguitar—Feb 14, 2023
@ZoeSchiffer "For You" is actually for him

CHAPTER 52

"These People Are Pillaging Us"

In early 2023, Twitter's daily revenue was at $5.93 million—down 40 percent year over year. Musk called a small group of employees together to discuss a fix: attaching a hefty price tag to Twitter's application programming interface, or API.

Twitter was already making roughly $134 million a year licensing its data to other companies, including Apple, Google, Yahoo Japan, and NTT Docomo. These were high-margin deals, but with interest payments on the $13 billion bank debt looming (January's payment alone was $300 million), Musk wanted to charge even more.

Employees thought it was unlikely that Musk could squeeze more money out of Apple or Google. What about the people who were accessing Twitter's data for free?

Historically, Twitter gave researchers free access to its API, allowing them to study trends in misinformation and hate speech. This research was critical to understanding the impact of social media. But it also resulted in bad headlines for Twitter. In 2021, an article in *The Guardian* read: "Twitter admits bias in algorithm for rightwing politicians and news outlets."

Some small developers got a similar deal, allowing them to build bots and apps. The projects weren't lucrative, but the idea that a developer could build a for-profit entity using Twitter's free data irked Musk.

"These people are pillaging us," he said, according to two employees who attended the meeting.

Musk wanted Twitter to shut down free access to Twitter's API—a move that would effectively kill research projects and public service bots that posted real-time updates about things like train schedules and earthquakes. (Later, Musk said that bots providing "good content that is free" wouldn't be completely cut off.)

Perhaps understanding how unpopular this decision would be, Musk justified it by claiming the API was being abused by scammers. "There's no verification process or cost, so easy to spin up 100k bots to do bad things," he tweeted.

The explanation made little sense to members of Twitter's trust and safety team. The free developer API did not include sign-up endpoints—it couldn't be used to create a Twitter account. While scammers sometimes leveraged the API to create automated posts, this only accounted for about 10 percent of the spam on Twitter.

When Twitter rolled out its new pricing, the tiers started at $42,000 a month for access to fifty million tweets and went up to $210,000 a month for access to two hundred million tweets. It was a prohibitively expensive price jump for the vast majority of researchers.

Twitter staffers created a spreadsheet of accounts that were going to have their access cut. Employees started from the bottom, targeting accounts that used the least amount of data and working their way up to the accounts that used the most. Each account was manually reviewed by a member of Musk's inner circle, usually James Musk or Christopher Stanley.

In the process, a number of "good" bots were blocked from accessing Twitter's API, including @SFBartalert, which sent automated messages about the Bay Area's public transportation service. When the account tweeted about having its API access severed and no longer being able to send notifications, a member of Twitter's development team reached out to apologize and reverse the suspension.

The API debacle showed just how far Musk was willing to go to stop people from, in his mind, taking advantage of Twitter. The project did not meaningfully move the needle on Twitter's revenue, at least in the short-term, according to an employee familiar with the project. Beyond that, cutting off access to researchers undoubtedly made Twitter less transparent. In May 2022, Musk tweeted "Sunlight is the best disinfectant." But in Musk's world, he alone was the sun.

CHAPTER 53

"Caught Red-Handed"

Elon Musk was growing increasingly frustrated with the information leaking out of Twitter. In December, the CEO had sent employees a warning:

"As evidenced by many detailed leaks of confidential Twitter information, a few people at our company continue to act in a manner contrary to the company's interests and in violation of their non-disclosure agreement (NDA)," he said. "This will be said only once: If you clearly and deliberately violate the NDA that you signed when joining Twitter, you accept liability to the full extent of the law & Twitter will immediately seek damages."

Despite the low morale, poor working conditions, and widespread layoffs, Musk couldn't understand why he didn't have more loyalty from his workforce. Perhaps threatening those who remained would make them more faithful.

It didn't. Someone leaked me this email, which I promptly tweeted. Then, the Friday after the Super Bowl, I got another scoop. Twitter was planning to unveil a new policy that only Blue subscribers would be able to use two-factor authentication, allowing them to secure their accounts by adding a phone number to the sign-in process. You didn't have to be a security expert to understand why protecting only paying users was a bad idea. Twitter users were pissed.

Musk ordered the security team to shut down Slack and find the leaker. Employees were told the app was down for "routine maintenance."

On February 22, 2023, Twitter had the equivalent of a snow day. Slack was down and Jira, a tool the company used to track feature updates and regulatory compliance, stopped working without an explanation.

Without a clear way to do work, a somber mood permeated the company. Few believed Slack was down for maintenance, whatever that meant.

Finally, on Friday, February 24, Randall Lin was sitting at his desk when he got a Google Chat message from a member of the corporate security team. They asked if he had time to speak that afternoon about a sensitive matter.

Lin said yes. He wasn't sure what it was about, but he wanted to help. The sooner Musk found and eliminated the problem, the sooner they could all get back to work.

When Lin got into the conference room, a member of the corporate security team was already waiting. Another man joined remotely via video conference. The men claimed they had proof that Lin was the source behind my article about Yang's firing and the one about Musk's tweets being boosted after the Super Bowl.

Lin was surprised—obviously, there had been a misunderstanding. He told them they had the wrong guy. "I've never talked to Zoë in my life," he said.

(This was true. Lin and I had never spoken. I'd heard his name from other sources, but always in the context of how close he was getting to the Goons, so I'd never even bothered reaching out, since I figured he wouldn't be willing to speak to me.)

"Have you ever screenshotted anything?" one of the men asked.

Lin thought. He'd once screenshotted an email from Parag Agrawal to show a friend how dumb and corporate it sounded. He decided it was best to be honest. "I screenshotted one email before Elon took over," he said.

The men kept pressing him about leaks, but Lin was adamant he'd never talked to me or Casey Newton. He told the men they needed to rule him out; he was trying to help them get to the bottom of this.

The men took his laptop for forensics and told him he could leave. "Can I get a replacement so I can do work?" Lin asked. The men said they would check and get back to him.

An hour later, Lin was sitting at his desk, scrolling on his phone, when a colleague asked him to step into the hallway. A security guard named Scotty was waiting. Lin's heart raced. Scotty was the person who walked employees out when they'd been fired.

"We can't get that laptop for you," Scotty said, according to Lin's recollection. "You should probably head home."

Lin noticed two other security guards waiting nearby. He went back to his desk and picked up his backpack. Scotty walked him to the elevator. As they rode down to the first floor, Lin said, "I tried to make sure I'd never see you like this." Scotty didn't respond.

An hour later, Lin's accounts were deactivated.

He met Alex, his girlfriend, for dinner, but he couldn't think about anything but work. This had to be a mistake. Surely, Musk would realize what had happened and bring him back. Lin texted his manager and told him what was going on.

"Yeah, I heard," the manager responded. "Let's see where the wind blows."

The next day, a Saturday, Lin heard from Twitter's HR team. They said Lin had violated the employee handbook. He was fired.

Once again, Lin texted his manager. "It's over," he said. The man replied with a melting monkey face emoji.

What the hell kind of response is that? Lin thought. He reached out to a member of the transition team and an employee on the security team, but neither responded. A colleague warned Lin that James Musk was telling people Lin had admitted to leaking dozens of articles—that he'd been caught red-handed. It wasn't clear whether James was lying or whether he was being lied to. Either way, Lin was collateral damage.

He had worked side by side with Elon and his lieutenants for months, listening to stories about their families, and pulling all-nighters at the office. He'd always known getting fired was a possibility. But experiencing it in real life hit different. Lin was crushed.

"The firing was so devastating to me because I did a lot of things to both appease Elon and shield my people and coworkers from him. The fact that they didn't do that for me sat really wrong with me," he said. "I realized that I was valued technically but that, somewhere along the way, someone lied about me, and my technical skill wasn't enough to save me."

CHAPTER 54

"I'm Not Going to Bullshit You"

The following day, on February 26, 2023, Twitter laid off two hundred people. The cuts represented roughly 10 percent of Twitter's workforce, including numerous product managers, data scientists, and engineers. JP Doherty's promotion was no longer optional. He was appointed the global director of TCC.

Doherty and Musk had discussed the position weeks before. "I need someone who knows what to do with Twitter's servers," Musk told him during their first one-on-one conversation. "I keep asking around and apparently that's you." Doherty acknowledged he had a good understanding of Twitter's hardware. He told Musk his nonnegotiable for taking the job. "I'm not going to bullshit you," he said. "I'm not going to paint you a rosy picture when there's not one." Musk seemed to appreciate his candor.

During their conversations, Musk asked Doherty how many kids he had. When Doherty said he just had two, the CEO (who has fathered eleven children) looked unimpressed.

A number of founders who'd joined Twitter when the company acquired their startups were laid off as part of the February purge. The most surprising of those cuts: Esther Crawford. Not only was she publicly, vocally loyal to Musk, her contract included accelerated vesting, meaning that Twitter owed her a sizable chunk of money when she left the company. "I made peace with the fact that I didn't have psychological safety at Twitter 2.0 and that meant I could be fired at any moment, and for no reason at all," she later said, adding that getting laid off was the best gift she'd ever received.

Other founders had their access to Twitter's internal systems cut, but they didn't receive an email about being terminated. Haraldur Thorleifsson, who spearheaded the launch of the edit button, tweeted at Musk days after the layoffs to get clarity about what had happened.

"Dear @elonmusk 👋," he tweeted. "9 days ago the access to my work computer was cut, along with about 200 other Twitter employees. However your head of HR is not able to confirm if I am an employee or not. You've not answered my emails. Maybe if enough people retweet you'll answer me here?"

"What work have you been doing?" Musk responded.

Thorleifsson said he'd need to break confidentiality to speak about his projects. Musk replied, "It's approved, you go ahead." Thorleifsson listed out his accomplishments: "Led the effort to save about $500k on one SaaS contract. Supported closing down many others," and "Led prioritization of design projects across the company to make sure we were able to deliver with a small team."

Then Musk told his 129 million followers that the Icelandic designer, who used a wheelchair due to muscular dystrophy, "did no actual work, claimed as his excuse that he had a disability that prevented him from typing, yet was simultaneously tweeting up a storm."

A former colleague responded to defend Thorleifsson. "As someone who has worked directly with @iamharaldur during a turnaround, this is super disappointing to see," the colleague wrote. "Not only is his work ethic next level, his talent and humility are world class. Exactly the kind of person you want on your team when the odds are stacked. I feel certain there's a deep misunderstanding somewhere in here of 'did no actual work.'"

The comment seemed to hit a nerve. Musk jumped on a video call with Thorleifsson. He admitted he'd been misled about the man's accomplishments.

"I would like to apologize to Halli for my misunderstanding of his situation," he tweeted. "It was based on things I was told that were untrue or, in some cases, true, but not meaningful."

The apology was too little, too late for JP Doherty. How could he work for someone who treated people with disabilities with such cruelty? A month before, his son, Rhys, had undergone his long-awaited eye surgery. For years, Doherty had helped Rhys go up and down the stairs in their two-story house in Alameda. His son would grab the banister with one hand and grip Doherty's arm in the other. Now, with the operation behind him, Rhys might be able to walk on his own, and Doherty was free to look for a new job. He wanted to resign immediately, in protest over how Musk treated people who had disabilities. But he had to keep working for Musk, for his family.

CHAPTER 55

"Official Company Communication"

Employees who'd been laid off in November 2022 had been waiting months to get their separation agreements. On January 4, 2023, the "nonworking notice period" came to an end. They were no longer Twitter employees. And yet, no documentation arrived.

Finally, on January 7, some employees started receiving emails from CPT Group, an administrative firm that wasn't accredited by the Better Business Bureau. The email instructed employees to click a link that redirected to www.usseparation.com.

I asked a few sources to check their inboxes to see if they'd received the email. None had. That was odd—I was already seeing news of it circulating on Twitter. Then I asked a source to check their junk folder. The email was there. Gmail had marked it as spam.

Twitter sent out a second memo. This time, the company clarified that it had engaged CPT Group to handle the separation agreements. The original email was an "official company communication and not a phishing attempt," Twitter said.

Twitter acknowledged that it was late sending out the separation agreements. But it didn't take responsibility for the delay. The company said the lawsuits that former employees had filed—many alleging wrongful termination and breach of contract—were to blame. "Please note that we would have sent you these agreements sooner, but a court order obtained by plaintiffs' attorneys resulted in this delay," the company said.

Twitter employees were offered one pitiful month of severance pay. In contrast, Meta employees in the US who were laid off in November 2022 received sixteen weeks' pay plus an additional two weeks' pay for every year of employment. Still, the economy was hurting, and jobs were scarce. Many signed the separation agreement, forfeiting the right to sue the company or speak about it publicly. As of October 2023, many employees from different waves of layoffs still had not received severance. Approximately two thousand former Twitter employees have attempted to pursue arbitration claims against the company, according to a 2023 court filing.

CHAPTER 56

"A Recently Fired Twitter Employee"

On Monday, February 27, 2023, Twitter announced that Slack was coming back. The company had archived all the old channels, including #social-watercooler, which had once been the site of the internal resistance to Musk's regime.

I was working on a follow-up story when I got a message on Signal from an unknown number.

"Hi Zoe," the message read. "I'm a recently fired Twitter employee . . ."

I asked to FaceTime. On the other end of the phone was a dark-haired young man wearing round, clear-framed glasses. He seemed nervous, pacing on the sidewalk while we talked.

His name was Randall Lin.

Lin was terrified that Musk was going to try to sue him for stories he didn't leak. He wanted information. Who had lied about him to Musk? And why? It was an interesting story—one that showed the complicated game that anyone had to play if they wanted to exist in Musk's inner orbit.

But Lin didn't want to participate in a story. He wanted to know why people believed he was a leaker. I told him honestly that I didn't know. He asked if I knew any labor lawyers. I gave him the names of two attorneys who were representing Twitter employees in ongoing legal fights over severance. Then I wished him well. I didn't expect to hear from him again.

CHAPTER 57

"Good Thing I Fired Him"

It took Twitter until March 2023 to move out of the Sacramento data center completely. The project involved 234 people and resulted in 2,573 server racks being moved to Portland and Atlanta. It saved Twitter $117 million a year in leases, power, and network costs. But it cost the platform in terms of stability.

In February, Twitter experienced four major outages. Jane Manchun Wong, a talented security researcher, tweeted that her Like counts were fluctuating. "Synchronization lag between our Portland and Atlanta data centers," Musk replied. "Should be fixed now."

"In retrospect, the whole Sacramento shutdown was a mistake," Musk told his biographer, Walter Isaacson. "I was told we had redundancy across our data centers. What I wasn't told was that we had 70,000 hard-coded references to Sacramento. And there's still shit that's broken because of it." In reality, Musk hadn't bothered to ask.

On March 6, a bug temporarily broke all the links on Twitter. It was the sixth major outage that Twitter experienced in 2023 (by comparison, in all of 2022 Twitter experienced nine outages).

That day, if a user tried to click on a link, a scary-looking error message read: "{"errors":[{"message":"Your current API plan does not include access to this endpoint, please see https://developer.twitter.com/en/docs/twitter-api for more information","code":467}]}."

"Twitter is down," an engineer wrote frantically in Slack.

"The entire thing."

Musk pulled a small group of engineers into a meeting and demanded answers. One of Doherty's colleagues blamed the incident on an employee who'd left the company. Musk seemed pleased with that answer. "Good thing I fired him," he said.

What Doherty's colleague had told Musk was technically true. But it wasn't how Twitter used to operate. In the past, the engineering organization had followed a "blameless postmortem" model, meaning the team would try to figure out what had gone wrong without pinning it on anyone in particular. This put all the engineers on the same side and focused their energy on finding the root of the problem, rather than simply pointing a finger. Besides, if Musk really wanted to find someone to blame, all he needed to do was look in the mirror.

Twitter was still online, but the service was more glitchy than it had been in years. "It feels like Twitter is a mangy animal with rabies shambling around a public park, foaming at the mouth, having periodic spasms, and waiting to die," wrote *New York* magazine columnist John Paul Brammer in March 2023. "But it hasn't yet."

Later that month, a security researcher at HackerOne told Twitter that the company's source code had leaked online. This was bad. "There was a high chance of keys and other technical secrets in the code that could lead hackers to get access to our internal systems," a source told me.

A member of Twitter's security team looked into the incident and realized that, since January, at least three security researchers had reported the leak to Twitter, but their emails had been ignored. The company used a vendor called NCC to validate reports from HackerOne. The code had leaked onto the developer platform GitHub, and NCC often marked GitHub reports as invalid, since Twitter purposely open-sourced certain projects.

Technically, the security team was supposed to review the HackerOne emails to make sure NCC didn't miss anything. But in the chaos that had engulfed the company since October, a lot of those emails had gone unread. The code had leaked on January 3, 2023, two months after Musk laid off half the company, but Twitter didn't find out until March 24.

Twitter sent GitHub a DMCA takedown notice and asked the company to preserve information about who'd accessed the code.

"Please preserve and provide copies of any related upload/download/access history (and any contact info, IP addresses, or other session info related to same), and any associated logs related to this repo or any forks thereof, before removing all the infringing content from Github," the company wrote.

GitHub removed the code, but it had already been downloaded thousands of times. The team couldn't track where it was spreading.

Twitter subpoenaed GitHub to force it to reveal information about the leaker. To my knowledge, their identity is still unknown.

The incident seemed to confirm Musk's worst fears about Twitter. Since October he'd assumed employees were trying to sabotage him. Tweeps were treated with extreme suspicion. This paranoia underpinned many of his decisions—from technical mandates (freezing the code base) to management ones (not giving employees advance warning about the layoffs) to inviting outsiders in to meddle (Hotz's internship and the Twitter Files) to sparring publicly with his past and present employees on Twitter (Sasha Solomon, Halli Thorleifsson).

The lack of trust had looked irrational. But now, Musk appeared vindicated.

As the security team investigated the leak, they found that the code that leaked on GitHub matched the code base in early November. An employee who'd been laid off on November 4 had apparently downloaded

Twitter's code before their access was cut. The question now was simply: Who?

The answer to this question should have been relatively straightforward. Who had accessed Twitter's code base prior to the layoffs and downloaded vast swaths of company data? The security team used two tools, Carbon Black and Uptycs, which it configured to track this kind of activity. If anyone had inserted a USB drive into their work computers—a clear indication they meant to download data—the security team was supposed to get an alert.

When the head of the team looked at the activity logs, however, they found that the tools had been implemented incorrectly. It's "not sabotage, they were inept," an employee tells me. No alert was ever sent. There wasn't one employee who'd inserted a USB drive into their work computer—there were a dozen.

Mohammed Qazi, who led security after the acquisition, now had to report the incident to regulators, including the FTC. When Musk suggested the company not immediately report it, Qazi resigned.

CHAPTER 58

"Inverse Startup"

On March 24, 2023, Musk sent a company-wide email estimating that Twitter was worth $20 billion, less than half of what he'd bought it for in October. Twitter was changing so fast it "can be thought of as an inverse startup," Musk said, noting the changes were necessary to ensure Twitter didn't go bankrupt. "At one point, we were only ~4 months away from running out of money!" He did not acknowledge that this existential threat was the result of him scaring off advertisers and saddling Twitter with $13 billion of debt.

"Post-Musk you have (1) a huge drop in advertising revenue caused by Musk quite intentionally alienating advertisers and users, (2) a tiiiiiiny offsetting bit of revenue from Twitter Blue that got comically small numbers of signups, and (3) like $300mm of *quarterly* interest expense. That is a much riskier business!" wrote *Bloomberg* finance commentator Matt Levine in an email to me, describing Twitter's financial situation in early 2023. "Offsetting that he fired everyone and didn't pay severance so maybe the numbers work, but a lot of the risk was self-created."

Musk told employees that he wanted to align their individual financial incentives with the company. To that end, every six months starting in 2024, Twitter would aim to have liquidity events so employees could sell stock on the secondary market. To Musk, this "achieved the public company advantage of having a liquid stock, but without the stock price chaos and lawsuit burdens of a public company."

In the email, Musk told employees: "I see a clear, but difficult, path to a >$250B valuation, which would mean stock granted now would be worth ten times more in the future."

Employees didn't have faith that Musk would follow through with the payments, since, among other things, he'd laid off much of the payroll staff. In March, Twitter sent some former staffers an email telling them the company had likely messed up their W-2s. "During a recent audit, we discovered that the ESPP (Employee Stock Purchase Plan) disposition that occurred in October 2022 was not recorded," the company wrote. "As a result, employees who purchased shares through our ESPP program and sold those shares in October 2022 may have received incorrect W-2 forms for the 2022 tax year."

Doherty was offered hundreds of thousands of dollars' worth of stock that would take four years to fully vest. "The kind of money they offered me would certainly solve a lot of problems," he said. "But it's not worth having a terrible feeling about what is going on, all the time." He continued looking for a new job, jokingly referring to the stock grants as "Spacebucks."

Arguably the most concrete part of Musk's email was that it revealed Twitter had a new name, one that brought him a step closer to fulfilling his long-held dream of X. "Going forward . . . all incremental compensation will be in the form of X Corp (Twitter) stock," he wrote. The following month, a filing in a frivolous lawsuit brought by alt-right activist Laura Loomer confirmed it: "Twitter, Inc. has been merged into X Corp. and no longer exists."

Musk tried to take down the massive TWITTER sign that hung outside the office but claimed the landlord would not allow him to do so. Instead, he ordered the *W* to be painted white, so it blended into the background. The sign now read T ITTER.

A week after Musk sent the email, Twitter posted its recommendation algorithm on GitHub, allowing outsiders to see how tweets were ranked. Unlike Twitter's source code, the recommendation algorithm didn't contain information that could easily be exploited.

Engineers had spent weeks cleaning up the code base in preparation for its public debut. "No doubt, many embarrassing issues will be discovered, but we will fix them fast!" Musk tweeted. When it was published, the person it seemed to embarrass most was him, as the code revealed that the algorithm specifically labeled whether a tweet was coming from Elon Musk. Other labels included Democrat, Republican, and "power user." Around this same time, I published a story revealing Twitter was allowing a select group of VIP users to "bypass heuristics that might otherwise limit how often their tweets are shown on the For You tab." Lin's list of canaries seemed to have taken on a new life since he left the company.

Seeing the recommendation algorithm go live was surreal for Lin, who'd written much of it himself. Later that month, as I considered whether to write this book, I called him back and pitched him on being a part of it. To my genuine surprise, he agreed. "I'm not vindictive in any way," he told me during one of our early conversations. Lin stressed that he only wanted one thing: to clear his name.

CHAPTER 59

"You're Welcome Namaste"

Even before the deal closed Musk was entranced with the idea of shifting Twitter's business model away from advertising and toward subscriptions. The move fit neatly into his worldview—that advertisers sucked, that Twitter's legacy verification system was biased, that Twitter's former leadership had slept on obvious revenue opportunities. The details of getting to a $250 billion valuation were hazy. But Musk wanted Twitter Blue to play a significant role. In November 2022, he had told employees he wanted half of Twitter's revenue to come from subscriptions.

Of course, the rollout of Musk's marquee subscription tier hadn't gone as planned. First, there was the significant issue with impersonators. Then the trust and safety team realized Apple didn't give the company private information about iOS subscribers, making it difficult to stop bad actors. Finally, very, *very* few people had signed up. After pausing the rollout in November, Musk relaunched Twitter Blue on December 12. Two months later, just 290,000 users had subscribed. "All together, the global number of subscribers would equate to around $28 million in annual revenue—less than 1% of the $3 billion Musk has said Twitter aims to make in revenue this year," read a detailed analysis in *The Information.* The $3 billion Musk hoped to make was still lower than Twitter's annual revenue in 2020, 2021, and 2022.

Verified organizations were by far the biggest disappointment. As part of the Twitter Blue relaunch, Twitter rolled out a portal where brands could apply to be verified. The application included a nonrefundable one-thousand-dollar fee. If a brand was approved, it got a gold check mark, indicating that it was the real deal.

By March 2023, only seventy-one brands were paying customers. The company rolled out invoicing, allowing organizations to get verified before making payment. Spammers who had no intention of paying flooded the pipeline. Twitter had to temporarily roll back the feature.

Luckily, Musk had a plan. On March 23, Twitter announced that it would soon take away the blue check marks belonging to users who didn't subscribe to Twitter Blue. Moving forward, Blue subscribers would be given preferential treatment in the algorithm, meaning tweets from their accounts would be more likely to be recommended.

Celebrities were quick to denounce the move. LeBron James told his 52 million followers he would not be subscribing. "Welp guess my blue ✔️ will be gone soon cause if you know me I ain't paying the 5 🤷🏾‍♂️," he tweeted (Twitter Blue was actually $8). Halle Berry told her followers she would also be losing the blue check, tweeting out a video of herself walking onto a stage with the caption, "Me joining you all tomorrow unverified 🥰."

Nevertheless, on April 20, 2023, Twitter started taking away legacy check marks from celebrities and journalists alike. The trust and safety team experienced *Groundhog Day*–level chaos as verified accounts impersonated the pope, a Sudanese paramilitary group, and Robert De Niro, among others.

The iconic blue badge took on an entirely new meaning overnight. Prior to April 20, the check mark signified that, if an account was verified, it belonged to the real Taylor Swift, the one and only Chrissy Teigen. For people who weren't household names, the badge had been a bona fide status symbol. Now, it was a tacit endorsement of Elon Musk.

Those who'd somehow retained their verification badges were quick to point out that they were *not* paying for Twitter Blue. Dril, the king of Weird Twitter, promoted a #BlockTheBlueChecks campaign. The blue check had become a badge of dishonor.

The change was so stark that Bobby Allyn, a journalist at NPR who'd subscribed to Twitter Blue, canceled his subscription and asked Musk to take away his badge. "Hey @elonmusk, I stopped paying for Twitter Blue," he tweeted on April 20. "Mind stripping off my blue check now? It's cramping my style on here."

Stephen King found that he, too, was still verified. Except, unlike Allyn, he'd never paid for a subscription.

"My Twitter account says I've subscribed to Twitter Blue. I haven't," he tweeted.

Musk responded: "You're welcome namaste 🙏."

Musk had given King, LeBron James, William Shatner, and a number of other high-follower accounts complimentary badges.

The great debadging of April 20 did not create the financial boost that Musk had hoped. The move drove a high number of sign-ups, but also a high number of cancellations. "Just before the purge yesterday, 19,469 of the 407k legacy verified accounts I had identified in early April had Twitter Blue," tweeted Travis Brown, a programmer and former Twitter employee, who analyzed the company's data. "Today that number for those same accounts is 19,497." The net increase, according to Brown, was just twenty-eight new Twitter Blue subscribers, providing $224 in new revenue per month.

CHAPTER 60

"Try It, but Don't Trust It"

In mid-April 2023, Elon Musk sat down for an on-camera interview with Fox News host and conservative firebrand Tucker Carlson, to discuss (among other things) the Twitter Files. "The degree to which various government agencies effectively had full access to everything that was going on at Twitter blew my mind, I was not aware of that," Musk said seriously. "Would that include people's DMs?" Carlson asked, his brow furrowed. "Uh, yes," Musk responded. Carlson couldn't hide his smirk. "Yes, because the DMs are not encrypted," Musk added.

The question let Musk announce an exciting project: Twitter was getting ready to launch encrypted DMs, a project that had failed to get off the ground in Twitter 1.0.

Popular messaging apps like WhatsApp and Signal brought end-to-end encryption to the mainstream, allowing users to talk to one another without fear of their messages being read by a third party. Even if the companies were subpoenaed by law enforcement, they couldn't decrypt users' messages. Ironically, this meant that the user didn't *have* to trust the platforms, they just had to trust the technology.

In 2016, Edward Snowden had pitched Jack Dorsey on launching encrypted DMs at Twitter. Engineers worked on the project for more than three years, going as far as to test it out internally, but it never launched, according to an internal report titled "What happened to encrypted DMs?" It took Twitter a year to get a license for Signal's technology to build the feature. Then the team working on encrypted DMs was reorganized, and the project was effectively abandoned.

Dorsey's failure was a boon to Musk, who could launch encrypted DMs using parts of Twitter's code from the first attempt, bolstering his claim that Twitter 2.0 was far more efficient than its predecessor.

"We want to enable users to be able to communicate without being concerned about their privacy, [or] without being concerned about a data breach at Twitter causing all of their DMs to hit the web, or think that maybe someone at Twitter could be spying on their DMs," Musk told employees during a meeting.

Security researcher Matthew Garrett was initially excited about Musk's announcement. Then he looked at the documentation Twitter released detailing what it planned to build, and he started to get concerned.

Twitter's plan was to have each user generate a pair of cryptographic keys, one public and one private. The public key would be uploaded to Twitter and associated with the user's account. If the user wanted to send an encrypted message, they would have to effectively ask Twitter for the private key associated with the user they wanted to talk to, and use that key to decode the message.

The problem, Garrett argued, was that a Twitter employee could add their own public key to the list of keys associated with another user, allowing them to obtain the corresponding private key and decrypt the message. A user's messages would be private from everyone—except Twitter. If the company was hacked, the messages could be compromised. And, if the government subpoenaed the company, it would be able to hand over someone's private messages.

Garrett pointed out these flaws on Twitter. Musk's lieutenant Christopher Stanley, who was leading the encrypted DMs project, responded and promised the company would publish a white paper about the process soon. "I had [cybersecurity firm] Trail of Bits audit our implementation. [CEO] Dan Guido and those folks are badass," he added.

But Twitter hadn't even signed a contract with Trail of Bits, as I learned when I reached out to a source at the social platform. Nearly every single person on Twitter's procurement team had been laid off, leaving the security firm without a clear point of contact.

Twitter launched encrypted direct messages for Twitter Blue subscribers on May 10, 2023, without a third-party review from Trail of Bits. In a blog post announcing the project, the company admitted that messages would not actually be end-to-end encrypted. "Currently, we do not offer protections against man-in-the-middle attacks," the company wrote. "As a result, if someone—for example, a malicious insider, or Twitter itself as a result of a compulsory legal process—were to compromise an encrypted conversation, neither the sender or receiver would know."

"Try it, but don't trust it yet," Musk tweeted. He promised the feature would get more sophisticated over time.

To Musk's fans, it was yet another example of the CEO's speed and commitment to transparency. To those who'd helped build Twitter's security apparatus, it was confounding. "Twitter folks, seriously, I left some design docs somewhere," wrote Lea Kissner, Twitter's former chief information security officer, on Bluesky. "Please use them."

CHAPTER 61

"Did Your Brain Fall Out of Your Head?"

Before Musk bought Twitter, the company's commitment to free speech had an important caveat. To operate in countries like Russia and Turkey, Twitter had to comply with local laws, even when those laws flew in the face of its stated values.

This tension was not unique to Twitter. Meta, Apple, Netflix, and Reddit all walked tightropes in countries with authoritarian governments, taking down content that clearly broke the law and pushing back on requests that were overly broad. The issue was far from straightforward. If the platforms failed to comply, they could be banned. Was it better to bend to the whims of an authoritarian government so the platform could continue operating, or to take a stand and risk users in that country losing access?

In some cases, the companies filed court orders, challenging takedown requests, as Twitter did in Turkey in 2014. Lumen, an independent research project from the Berkman Klein Center for Internet & Society at Harvard University, tracked the companies' progress, publishing reports about how well they protected speech.

Netflix had seen its commitment to artistic freedom repeatedly tested in Turkey, where President Recep Tayyip Erdoğan took issue with content starring LGBTQ characters, particularly when those characters lived in the Middle East. In 2020, the streaming giant canceled a Turkish Netflix original called *If Only*, rather than comply with the government's request to remove a gay character from the show.

In the six months that preceded Musk's takeover, Twitter received 550 requests from courts and governments, according to self-reported data compiled by Lumen. "These requests included orders to remove controversial posts, as well as demands that Twitter produce private data to identify anonymous accounts," *Rest of World* reported, noting that Twitter fully complied with about 50 percent.

In the six months after the takeover, the number of requests almost doubled to 971. Under Musk, Twitter was fully complying with a staggering 80 percent.

The increase in requests might be explained by countries like India and Turkey passing restrictive speech laws, which gave them more latitude to request takedowns. But those laws also applied to companies like Meta, which had seen a slower increase in requests and hadn't meaningfully upped its compliance.

Instead, it appeared that while Twitter was rolling back content moderation rules in the US, it was caving to authoritarian leaders more readily than it had under the leadership of Jack Dorsey, Parag Agrawal, and Vijaya Gadde.

First, in March 2023, the company complied with a court order to block the official Twitter account of Pakistan's government from being viewed in India.

Then, two months later, Twitter announced that it had restricted some posts in Turkey ahead of a contentious election between Erdoğan and his opponent Kemal Kiliçdaroğlu. The company didn't say which posts it blocked, but reports indicated that Erdoğan critics and at least one investigative reporter were among the accounts that had been impacted. (Erdoğan later won the election.)

Twitter users were quick to point out Musk's hypocrisy. "The Turkish government asked Twitter to censor its opponents right before an election and @elonmusk complied—should generate some interesting Twitter Files reporting," cultural and political commentator Matt Yglesias tweeted sarcastically to his half a million followers.

"Did your brain fall out of your head, Yglesias?" Musk snapped back. "The choice is have Twitter throttled in its entirety or limit access to some tweets. Which one do you want?"

Wikipedia cofounder Jimmy Wales provided an answer. "What Wikipedia did: we stood strong for our principles and fought to the Supreme Court of Turkey and won," he tweeted. "This is what it means to treat freedom of expression as a principle rather than a slogan."

CHAPTER 62

"I Need to Do Some Deleting"

As Elon Musk played nice with authoritarian governments, Twitter's reputation with US regulators was deteriorating. In March 2023, the FTC acknowledged it was investigating Twitter's privacy practices—an unusual move for the government agency, which rarely commented on ongoing probes. In a series of letters sent after Musk bought the company, the FTC demanded to see internal communications from the CEO, information about the launch of Twitter Blue, and details about the outside journalists who'd gotten access to internal documents as part of the Twitter Files.

Between November 10, 2022, and February 1, 2023, the FTC sent over a dozen demand letters to Twitter. According to the House Judiciary Committee, the requests included: "Every single internal communication 'relating to Elon Musk,' by any Twitter personnel—including communications sent or received by Musk—not limited by subject matter, since the day Musk bought the company," "When Twitter 'first conceived of the concept for Twitter Blue,'" and "Information disaggregated by 'each department, division, and/or team,' regardless of whether the work done by these units had anything to do with privacy or information security."

Twitter responded by asking a US District Court in San Francisco to end the consent order. "The Court should not permit the FTC to continue invoking judicial power to prosecute an investigation so infected by bias that it has lost any plausible connection to lawful purposes," it wrote.

House Republicans jumped in to defend Musk's company, arguing that the "FTC has been attempting to harass Twitter and pry into the company's decisions on matters outside of the FTC's mandate." The Justice Department urged the judge to deny this request, according to a document filed on behalf of the FTC.

Employees were growing increasingly concerned about the ongoing investigation. Twitter was fined $150 million the first time it violated the consent order. If the agency found that it had violated the agreement a second time, the fine would likely be much, much higher.

Project Eraser, the initiative to delete user data automatically and completely, still hadn't launched. Prior to Musk's takeover, the project was set to roll out in the fall of 2022. Then Musk laid off half the company, and Twitter's chief information security officer, Lea Kissner, resigned. Seven months later, it still wasn't done.

"Twitter is unable to automatically erase/scrub the entirety of a user's data after they submit a data deletion request, or after the retention period ends, or after the purpose for which the data was collected ends," an internal security assessment read. "However, data can still be deleted manually if necessary."

The company also hadn't implemented adequate safeguards against internal threats—a surprising omission given Musk's preoccupation with sabotage. "Employees have access to systems that store, process, and transmit personal data and there are no controls to prevent data abuse or misuse," the assessment read. "We do not know which employees have access to what data."

There was also a rising risk that Twitter was not following the flyway process and conducting thorough privacy reviews before launching new features, which was a key part of the FTC consent order. Damien Kieran, Twitter's former chief privacy officer, told the Department of Justice that the new Twitter Blue "was implemented so quickly that . . . the security and privacy review was not conducted in accordance with the company's process for software development."

In a section titled "rising risks," the internal security assessment echoed Kieran's claim: "Product and engineering teams plan, develop, and launch new products or changes to existing products' functionalities without review from legal, privacy, and information security teams, which could result in legal or regulatory fines and penalties and direct violation of the FTC Consent Order."

Even the shutdown of the Sacramento data center had alarmed FTC regulators. Per company policy, the machines were supposed to be wiped before they were moved from Sacramento to Portland or Atlanta. But, according to Twitter's former director of threat management, Seth Wilson, employees didn't have time. The Department of Justice alleged: "In fact, the relocated servers were not only unwiped, but they also contained . . ."

The rest of the sentence was redacted, leaving everyone to wonder what might have gone wrong.

Twitter's rising instability was also making it nearly impossible to ensure user data was secure.

Back in April, Twitter users had noticed that Circles, a feature that allowed people to share their thoughts with a small subset of followers, was broken. The whole point of Circles was to make it "easier to have more intimate conversations," according to a Twitter blog post. Launched during Parag Agrawal's short tenure, Circles had meaningfully increased the number of people who tweeted for the first time in years, as many weren't comfortable posting publicly. Now those conversations were showing up in the For You tab, as if Instagram had started blasting Close Friends stories on the Explore page.

"Confirmed someone I'm not even following was able to see a private

Twitter Circle tweet," wrote Twitter user @t3dotgg. "This hurts trust in the platform a lot. Should be top priority @TwitterEng."

"Holy fu ck I need to do some deleting," another user responded.

For months, Musk had been pushing engineers to simplify the Twitter code base. In the process, they'd deleted a backup filter to remove Circle tweets that should not be visible to non-Circle members.

In total, the bug impacted 780,481 private tweets.

It took a month for the company to acknowledge the incident to users who'd been impacted. By then, regulators were already on alert.

CHAPTER 63

"The Same Rules & Rewards"

Even as the FTC deepened its investigation into Musk's Twitter, the platform was becoming more fervently embraced by Republican pundits and lawmakers. In April 2023, Fox News ousted Tucker Carlson, and the disgraced host started looking for a new home for his show. The backdrop of the dismissal was a 2021 defamation lawsuit brought by Dominion Voting Systems, which accused Fox of peddling an election fraud conspiracy that claimed Dominion's machines switched Trump votes in favor of Biden. Fox paid Dominion $787 million—the largest-known settlement for a defamation case in US history, and Carlson's dismissal was reportedly a condition of the agreement.

A month after his firing, on May 9, 2023, Carlson announced he was taking his show to Twitter. The platform was reshaping itself into a right-wing social network. "Twitter is essentially following the playbook of platforms like Rumble, which used to be the go-tos for canceled and deplatformed right-wingers seeking a soft landing and the promise of revenue," explained Charlie Warzel in *The Atlantic*. A month earlier, Parler, which billed itself as an "uncancelable" free speech alternative to Twitter, shut down. "No reasonable person believes that a Twitter clone just for conservatives is a viable business any more," its parent company said.

Carlson flattered Musk in his announcement. "Amazingly, as of tonight, there aren't many platforms left that allow free speech. The last big one remaining in the world, the only one, is Twitter, where we are now. Twitter has long served as the place where our national conversation incubates and develops," Carlson said, wearing a gingham shirt far more casual than the buttoned-up suit and tie he wore on Fox News. If Carlson expected a warm welcome from Musk, he might have been surprised when the CEO responded to his announcement by clarifying Twitter hadn't signed a contract with him. "On this platform, unlike the one-way street of broadcast, people are able to interact, critique and refute whatever is said," Musk tweeted, before adding that Carlson would be "subject to the same rules & rewards of all content creators." Musk then urged left-wing content creators to join Twitter.

Still, Carlson's embrace of the social platform did not go unnoticed by conservative politicians—particularly those who felt burned by the mainstream press and aligned with Musk's antiwoke ideology.

In May, Florida governor Ron DeSantis announced he was running for president on Twitter Spaces. Months earlier, he'd attended a dinner hosted by David Sacks, Musk's longtime collaborator. Sacks, a member of Twitter's Goon squad, was well on his way to becoming a political powerhouse, leveraging his network and chart-topping podcast to fundraise for Republican candidates.

The move might've been a win for Musk, but it proved a problem for the Twitter Spaces team, which previously had around a hundred employees and now had just three.

Spaces technology relied on servers that Twitter rented from Amazon Web Services (AWS). Unfortunately, these servers were "insanely underprovisioned," according to an employee who'd worked on them. In the past, the technology had been able to autoscale, adding more capacity as listeners entered the room. But the process was bound to buckle under pressure. JP Doherty encouraged the team to run some tests, but they didn't have time. A colleague assured him that AWS would be able to handle the spike in traffic.

The day of the event arrived and Musk, Sacks, and DeSantis hyped it up on Twitter until hundreds of thousands of people were clamoring to get into the room.

From the moment it started, the Twitter Space was plagued with technical problems. The first room, hosted by Musk, crashed, because so many people were trying to get in. The second room, hosted by Sacks, started with an extended period of silence.

"Now it's quiet," a voice finally whispered. "We got so many people here that we are kind of melting the servers," David Sacks said. Musk, apparently unaware that his microphone was on, admitted, "servers are straining somewhat."

When the conversation finally got started, only around 300,000 users listened in. (In contrast, a 2016 Facebook Live event where *BuzzFeed* employees put rubber bands around a watermelon until it exploded got more than 800,000 concurrent viewers.)

After the event, Sacks and Musk hopped on a call with DeSantis's campaign manager, Generra Peck, to explain what happened. They said that nearly one million people had been trying to get into the room.

Internal documents showed that, out of every one hundred people who tried to get into Musk's Twitter Space, as few as seventy-one had been able to do so, due to technical failures. The team resolved the problem by having Sacks host the next room.

If Musk and Sacks shared that with Peck, she didn't say. By the next day, the campaign was spinning the event as a smashing success, sending fundraising emails that said DeSantis "breaks systems," including the internet and corporate media.

CHAPTER 64

"Every Parent Should Watch This"

Jeremy Boreing, co-CEO of conservative news site *The Daily Wire*, was sick of the hypocrisy he saw in Silicon Valley. For years, he'd been locked in a battle with the big tech platforms, fighting to share a conservative message on platforms that seemed bent on censorship.

"Tech giants like @youtube, @facebook, @tiktok_us, and—until @elonmusk wrote a check—@Twitter, flagging, banning, throttling, shadow banning, and demonetizing us is a constant occurrence," Boreing claimed in a tweet. It didn't matter that these were private platforms enforcing their own policies around misinformation and hate speech. To conservatives like Boreing, it was censorship to punish *The Daily Wire*'s reporters for violations ranging from misgendering a trans woman to spreading conspiracies about vaccines.

Boreing sensed that he might have an ally in Twitter's new owner. Musk had repeatedly spoken about his commitment to free speech at Twitter, of course, and seemed increasingly hostile to woke politics. Whatever Twitter did in countries like Turkey was its business. Boreing was focused on the United States. It didn't hurt that Musk was doubling down on audio and long-form video—two formats that were key to *The Daily Wire*'s success.

In May, Boreing announced that *The Daily Wire* would release all its podcasts for free on Twitter. "At this moment, Twitter is the largest free speech platform in the world," he wrote.

Around this time, *The Daily Wire* approached Twitter with another opportunity: distributing its 2022 documentary *What Is a Woman?* starring right-wing activist Matt Walsh. To Walsh and his supporters, the film was an "unflinching" look at "how a word as mundane and easily defined as 'woman' has become a pride flag draped hot potato." To most people, it was transphobic propaganda. According to critic Jessie Earl, who is trans, it was "a film made to sermonize to the already converted, those already primed to see the simple existence of trans people as worthy of laughter and contempt rather than curiosity or empathy, then hand them weapons in the form of lazy disinformation and hate."

The film hadn't gotten much attention when it came out. Now, a year after its release, Boreing saw a chance to change that. In a lengthy tweet thread on June 1, 2023, Boreing laid out what happened next:

Initially, Twitter responded to *The Daily Wire*'s pitch with enthusiasm. The company offered *The Daily Wire* a package to host the movie on a dedicated event page and promote it to every Twitter user for ten hours, Boreing claimed. *The Daily Wire* accepted, and the two parties signed an agreement.

There was only one problem: it seemed no one at Twitter had seen the film. The company asked for a screener to "better understand what parts may 'trigger' users, so they could better prepare their response," Boreing said. *The Daily Wire* sent the company a copy.

After employees watched the film, they determined it violated Twitter's hateful conduct policy. A representative from the company told *The Daily Wire* that—far from promoting the film—Twitter would be limiting its reach, according to Boreing.

In Boreing's account, the company called out two instances of misgendering that violated its rules: a father referring to his fourteen-year-old trans son as "her," and a store owner misgendering a trans customer.

Boreing was incensed. In April, Twitter had quietly deleted a line from its hateful conduct policy that banned "targeted misgendering or deadnaming of transgender individuals." I reported on the change at the time, noting the trust and safety team hadn't been told it was happening, to which Ella Irwin tweeted: "My team was not told this because it is actually not true. To be clear, literally no one at Twitter has said we will stop protecting trans users from abuse and harassment." Her carefully worded response did not address the altered policy language. Now, it appeared Boreing had made the same assumption I had: in taking out the policy specifically banning misgendering, Twitter was opening the door to allow more content on the platform.

Boreing claimed that Twitter told him it had removed the line because it believed misgendering was covered under the broader harassment policy. Surprising, given that Musk had made it abundantly clear to employees that he objected to Twitter's policies against misgendering ("Elon doesn't believe in misgendering policies," a former trust and safety employee tells me).

Nevertheless, if *The Daily Wire* wanted Twitter to promote the film, it would need to edit out the violative scenes, Boreing said. If it didn't, the film would be suppressed, and *The Daily Wire*'s followers wouldn't see it on their feeds. Anyone who wanted to see the movie would have to find a direct link to it, or navigate there from *The Daily Wire*'s account page.

"Of course, saying 'you have the right to speak, but we'll make sure no one hears you' is a bit like saying 'you have the right to cast a vote, but we'll make sure it isn't counted.' That's not a right at all!" Boreing tweeted.

The Daily Wire was well versed in using content-moderation disputes to its advantage. "@elonmusk is not beholden to conservatives. He has the right to run his business as he sees fit. But if Twitter is going to throttle one side of one of the most important debates facing society, it cannot claim to champion free speech," Boreing tweeted. He announced that *The Daily Wire* would be posting the full film on Twitter on June 1 as planned. "Will Twitter make good on their threat to throttle it and label it 'hateful conduct,' or will Twitter live up to its great promise?" he asked his followers, in a tweet that was viewed 582,100 times.

"This was a mistake by many people at Twitter," Musk responded. "[The film] is definitely allowed."

By the time the movie was uploaded to Twitter, the Streisand Effect was in full swing, as the controversy became national news. Twitter applied visibility filtering to stop the film from going viral—but that seemed to only make it more popular. The phrase "What is a woman" began trending on Twitter. Within days, *The Daily Wire*'s tweet that embedded the full film had 170 million views. (*The Daily Wire* compared this with the 27.3 million people who'd watched President Biden's State of the Union in January, although it was not clear that the number of people who'd watched the movie was anywhere close to the number of people who'd seen the tweet.)

"Every parent should watch this," Musk, the parent of a trans woman, said in a quote-tweet of the film.

On June 1, 2023, the day the film was uploaded to Twitter, Ella Irwin resigned. She did not share the reason publicly, though *The Daily Wire* speculated that she was responsible for labeling the film as a possible hateful conduct violation.

"So one or two people noticed that I left Twitter yesterday," she tweeted on June 2. "I know there's been a lot of speculation regarding what happened. Was I fired? Did I quit? Why?? Here's what really happened: 🧵 1/24." That turned out to be a joke.

"Just kidding folks. 😂," Irwin said in a follow-up tweet. "There's no thread."

She'd learned a thing or two from Elon Musk.

CHAPTER 65

"Someone Foolish Enough to Take the Job"

Linda Yaccarino, head of advertising at NBCU, had long wanted to be a chief executive. In the industry, she was known as the "Velvet Hammer," a savvy dealmaker who wined and dined potential advertisers and inked high-profile deals. Yaccarino was a vocal Trump supporter, according to two close associates; in 2018 Trump appointed her to serve on his administration's Council on Sports, Fitness & Nutrition.

In December, Musk had started hinting that Twitter might be in the market for a new chief executive. Tesla shareholders were angry the share price had fallen 28 percent in the month and a half since Musk bought Twitter. "Wake up tesla [board of directors]—what is the plan? Who is running tesla and when is Elon coming back?" tweeted investor Ross Gerber, who owned $74 million worth of Tesla stock.

That month, Musk polled his Twitter followers to ask whether he should step down as CEO. The majority of respondents (57.5 percent) said yes. One notable exception was Robert F. Kennedy Jr., an environmental lawyer and vocal antivax advocate who'd go on to run for president. "No! Do not step down!" he tweeted at Musk. "Stand up for free speech. The First Amendment needs you." Musk said he'd resign "as soon as I find someone foolish enough to take the job!"

Yaccarino invited Musk to join her in Miami for MMA Global's POSSIBLE conference in April, a major event in the advertising industry. Initially,

Musk agreed. Then SpaceX decided to launch Starship, its Mars rocket, around the time of the event. "He was going to blow it off, obviously," Musk's biographer, Walter Isaacson, tells me. At the last minute, SpaceX postponed the launch due to a technical problem, freeing up Musk's schedule to attend the conference. He decided to go. Antonio Gracias, his friend and adviser, lived in Miami. "He was like, 'OK, I'll hang with Antonio and I'll go to this conference,'" Isaacson says.

The afternoon of April 18, 2023, Yaccarino sat onstage in bright yellow stilettos that matched her chic skirt suit. She was beaming. She'd successfully snagged Musk for the event, and he'd even offered to bring his two-year-old son onstage. Then, at the last minute, X Æ A-12 decided he didn't want to go.

"I thought X was coming out," Yaccarino said in a heavy New York accent. "I had some extra questions for him."

"I think he changed his mind," Musk said.

Yaccarino laughed like he'd said something hysterical.

"Many of you in this room know me, and you know I pride myself on my work ethic," she said, after a flattering introduction of Musk. "But, buddy, I've met my match."

The two talked at length about Musk's vision for Twitter. After the conference, they went to dinner with a group of top-tier advertisers and ended up chatting until midnight, according to Isaacson.

Yaccarino pitched Musk on giving her the job of CEO. It made sense—she respected him, had deep relationships with blue-chip advertisers, and could put a friendly and professional face on the business.

Musk saw the appeal. Despite his misgivings about the ad industry, it was clear Twitter couldn't survive on subscriptions. By May 2023, internal documents showed Twitter had just 535,000 monthly Blue subscribers. (As a point of comparison, when Snap launched its Snapchat+ premium service, it hit one million subscribers in two months.)

In the days following the advertising conference, Musk was distracted by the SpaceX launch, which had been rescheduled to April 20. (The Starship rocket exploded shortly after launch, spewing rubble over miles of land.) Yaccarino grew nervous that Musk wasn't returning her calls. She called Isaacson to ask for advice, which he did not feel comfortable giving. Instead, he connected her to Jared Birchall.

Finally, in early May, Musk made up his mind. Yaccarino would become CEO and Musk would stay on as executive chair and chief technology officer. Yaccarino accepted but said she'd need a few weeks to wind down her work at NBCU.

Then, on May 11, Musk blindsided her with a tweet: "Excited to announce that I've a new CEO for X/Twitter," he wrote. "She will be starting in ~6 weeks!" Kara Swisher soon broke the news that the "she" in question was most likely Linda Yaccarino.

The news had an immediate impact on Twitter's reputation with the advertising industry. GroupM, the top-tier advertising firm that had previously rated Twitter as a high-risk ad buy for clients, said that it no longer considered the company high-risk and was "cautiously optimistic" about Yaccarino.

In many ways, Yaccarino's appointment saw Twitter coming full circle. Musk had come in with big ideas about how to change the platform, but in the end, Twitter still needed advertisers. As long as that stayed true, the platform would continue to walk a tightrope, trying to balance free speech with brand safety.

If there was one area where Yaccarino could immediately make her mark, it was in encouraging Musk to pay his bills. Musk might be willing to alienate everyone in the tech industry who wasn't immediately useful to him, but Yaccarino didn't have that luxury. Neither did Twitter, for that matter.

At the time, Twitter owed Google more than $42 million for cloud computing services. Musk did not want to pay the bill. Damn the fact that Twitter needed Google more than Google needed Twitter. Damn the fact that the money was for a contract that Twitter had signed and for services that Google had already rendered. Twitter's own employees had told Musk that the contract was poorly negotiated, feeding into his perception of the company's former leadership as financially irresponsible. He wasn't going to pay for their mistakes—not reputationally, and certainly not fiscally.

Twitter signed the multiyear $1 billion cloud deal with Google in 2018. In 2022, Twitter spent $261 million on Google Cloud services, according to internal documents. It projected it would spend $381 million in 2023.

Many believed Twitter had committed to buying more capacity than it needed from Google. But it couldn't just walk away. First, Google was one of Twitter's top advertisers, spending roughly $200 million a year, according to an employee with visibility into the deal. Second, Google was one of Twitter's largest firehose clients, paying $150 million across a two-year contract to be able to show tweets in search results. Third, Twitter was dependent on Google Cloud for much of its trust and safety work. Tools like Smyte, which scanned Twitter to flag posts for content moderators, ran on Google Cloud servers. These tools were critical to the trust and safety team, which needed relatively little of the Google Cloud budget. Under Musk, trust and safety was spending a mere $1.2 million a year on GCP, 0.4 percent of Twitter's Google Cloud budget.

By March, Twitter had negotiated down its Google Cloud bill to $178 million for 2023. But then the company stopped paying its invoices. Musk reasoned that if Twitter could shift its work off Google Cloud, getting its spend south of $100 million, it could go to the negotiating table with more leverage and lower its costs even further.

The engineering team rushed to move all Twitter services off Google Cloud as quickly as possible. Employees on the trust and safety team were concerned. Critical parts of the child safety detection pipeline ran on Google Cloud. They were told Smyte wasn't going to be transferred.

By June, it was clear that work was going slower than expected. Early that month, Twitter had a meeting with Google to renegotiate its contract. The discussion did not go well. Google refused to budge until Twitter paid its outstanding invoices. After the meeting, Twitter employees briefed Musk and Yaccarino and asked them what to do. If Twitter didn't pay the bill, it risked having its Google Cloud services shut off before Twitter's services were transferred to another data center. The results would be catastrophic for trust and safety.

The two executives thought it over and decided to ask for an extension. They wanted to wait a week to decide whether or not to pay the bill. Surprisingly, Google accepted.

Finally, just before the deadline, Twitter agreed to pay. It appeared Yaccarino was doing her job, helping Musk and the company narrowly avert disaster. The Velvet Hammer had arrived.

CHAPTER 66

"Serious Numbers"

On June 9, Musk announced that Twitter would finally start sharing ad revenue with creators who subscribed to Twitter Blue, a feature he'd promised in February. The first payments would total $5 million.

The move was part of Musk's plan to lure creators back to Twitter. In the eight months since he'd bought the company, the platform had had a number of high-profile departures. Gigi Hadid deleted her Twitter account. Meek Mill said he was done "forever." Elton John stopped posting because of misinformation. The platform still had scores of celebrity users, but to compete with platforms like YouTube and TikTok, it needed to give creators monetization tools.

In April, Twitter relaunched a subscription service so users could pay for access to exclusive content, a feature that was previously called Super Follows. Shortly after the relaunch, Twitter hit a rate limit with Google that stopped new creators from turning on subscriptions if they used an Android device. Lex Fridman, a well-known podcast host, complained, and employees worried he'd escalate the issue directly to Musk. The engineering team emailed Google a number of times to ask for a limit increase. "Our external partners don't always work with the same sense of urgency that we have," a director told Musk in a Saturday morning email. He said the issue would be fixed by the following Monday.

"I was just about to email about this!" Musk responded. "Please tell the

Google person that this needs to be fixed now, not Monday, or I will be calling personally." (The issue was swiftly resolved.)

On June 29, Musk sent an email to Twitter's product team, asking when creators would start getting paid for ads and subscriptions in "serious numbers." He stressed that the project was a high priority.

An engineering director responded that subscription payments were already happening. Musk himself had received roughly $90,000 in subscription revenue in mid-June. "We have a number of other creators that received payments in the $10k–$20k range," the director said.

When the ad revenue started hitting creators' accounts, the total was lower than Twitter had initially said—just $1.3 million across 250 accounts, according to internal documents. The company left out creators who didn't have a Stripe payments account connected to their Twitter profile. Roughly a dozen payments failed because the creator had moved and their Stripe account was linked to a country where they no longer lived. "We're working directly with these creators to find a workaround to get them paid ASAP," a product manager explained.

Two creators were accidentally paid twice. Twitter decided it would deduct the extra payment from their future earnings.

Andrew Tate, a far-right influencer who was facing rape and human trafficking charges in Romania, said he received $20,000. Other creators, like the pseudonymous shitposter @catturd2, were left out of the first round of payments completely. After all the free content he'd made for Twitter, @catturd2 was pissed.

"Most, not all, of the hand-picked Twitter accounts who were given big checks, not because of their ad revenue potential, but for being teacher's pets are becoming insufferable," @catturd2 tweeted. "They used to have accounts that interested me but now it's just a 24 hour Twitter advertisement on their pages. If this is what you have to do to get paid for the thousands of advertisements on your page, you can count me out. Honestly, I tried to figure out how I could get paid at first because of the size of my engagements and the years of work I've put in, but now that I see what's going on, I'm good. I was totally ignored by everyone I asked about this, except one good person. I'll just stay shadow-banned and say what I want."

A product manager spun the launch positively to Musk. "Reception was very positive overall, with lots of buzz on and off platform," he wrote in an internal email.

The launch drove a modest increase in Twitter Blue sign-ups, from the daily average of five thousand to around seventy-five hundred, still a small fraction of Twitter's monetizable daily active users. Musk was pleased. "Nice work," he told the team. "A 50% increase is a big deal in signups!" He told employees to send a note to every account around the world that would have been eligible for payments if they subscribed to Twitter Blue. Musk might hate advertising, but with Twitter's finances still in free fall, he was ready to sell. Shortly after, I saw the alert show up on my own account.

CHAPTER 67

"Sorry, You Are Rate Limited"

While the finer points of running a social media business can be debated, one basic truth is that they all run on attention. Tech leaders are incentivized to grow their user bases so there are more people looking at more ads for more time. It's just good business.

As the owner of Twitter, Musk presumably shared that goal. But he claimed he hadn't bought Twitter to make money. This freed him up to focus on other passions: stopping rival tech companies from scraping Twitter's data without permission—even if it meant losing eyeballs on ads.

Data scraping was a known problem at Twitter. "Scraping was the open secret of Twitter data access. We knew about it. It was fine," Yoel Roth wrote on the Twitter-alternative Bluesky. AI firms in particular were notorious for gobbling up huge swaths of text to train large language models (LLMs). Now that those firms were worth a lot of money, the situation was far from fine in Musk's opinion.

In November 2022, OpenAI debuted ChatGPT, a chatbot that could generate convincingly human text. By January 2023, the app had over 100 million users, making it the fastest-growing consumer app of all time. Three months later, OpenAI secured another round of funding that closed at an astounding valuation of $29 billion, more than Twitter was worth, by Musk's estimation.

OpenAI was a sore subject for Musk, who'd been one of the original founders and a major donor before stepping down in 2018 over disagreements with the other founders. After ChatGPT launched, Musk made no secret of the fact that he disagreed with the guardrails that OpenAI put on the chatbot to stop it from relaying dangerous or insensitive information. "The danger of training AI to be woke—in other words, lie—is deadly," Musk said on December 16, 2022. He was toying with starting a competitor.

Near the end of June, Musk launched a two-part offensive to stop data scrapers, first directing Twitter employees to temporarily block "logged out view." The change would mean that only people with Twitter accounts could view tweets.

"Logged out view" had a complicated history at Twitter. It was rumored to have played a part in the Arab Spring, allowing dissidents to view tweets without having to create a Twitter account and risk compromising their anonymity. But it was also an easy access point for people who wanted to scrape Twitter data.

Once Twitter made the change, Google was temporarily blocked from crawling Twitter and serving up relevant tweets in search results—a move that could negatively impact Twitter's traffic. "We're aware that our ability to crawl Twitter.com has been limited, affecting our ability to display tweets and pages from the site in search results," Google spokesperson Lara Levin told *The Verge*. "Websites have control over whether crawlers can access their content." As engineers discussed possible workarounds on Slack, one wrote: "Surely this was expected when that decision was made?"

Then engineers detected an "explosion of logged in requests," according to internal Slack messages, indicating that data scrapers had simply logged into Twitter to continue scraping. Musk ordered the change to be reversed.

On July 1, 2023, Musk launched part two of the offensive. Suddenly, if a user scrolled for just a few minutes, an error message popped up. "Sorry, you are rate limited," the message read. "Please wait a few moments then try again."

Rate limiting is a strategy that tech companies use to constrain network traffic by putting a cap on the number of times a user can perform a specific action within a given time frame (a mouthful, I know). It's often used to stop bad actors from trying to hack into people's accounts. If a user tries an incorrect password too many times, they see an error message and are told to come back later. The cost of doing this to someone who's forgotten their password is low (most people stay logged in), while the benefit to users is very high (it prevents many people's accounts from getting compromised).

Except, that wasn't what Musk had done. The rate limit that he ordered Twitter to roll out on July 1 was an API limit, meaning Twitter had capped the number of times users could refresh Twitter to look for new tweets and see ads. Rather than constrain users from performing a specific action, Twitter had limited *all* user actions. "I realize these are draconian rules," a Twitter engineer wrote on Slack. "They are temporary. We will reevaluate the situation tomorrow."

At first, Blue subscribers could see six thousand posts a day, while nonsubscribers could see six hundred (enough for just a few minutes of scrolling), and new nonsubscriber accounts could see just three hundred. As people started hitting the limits, #TwitterDown started trending on, well, Twitter. "this sucks dude you gotta 10x each of these numbers," wrote user @tszzl.

The impact quickly became obvious. Companies that used Twitter direct messages as a customer service tool were unable to communicate with clients. Major creators were blocked from promoting tweets, putting Musk's wish to stop data scrapers at odds with his initiative to make Twitter more creator-friendly. And Twitter's own trust and safety team was suddenly stopped from seeing violative tweets.

Engineers posted frantic updates in Slack. "FYI some large creators complaining because rate limit affecting paid subscription posts," one said.

Christopher Stanley, the head of information security, wrote with dismay that rate limits could apply to people refreshing the app to get news about a mass shooting or a major weather event. "The idea here is to stop scrapers, not prevent people from obtaining safety information," he wrote.

Twitter soon raised the limits to ten thousand (for Blue subscribers), one thousand (for nonsubscribers), and five hundred (for new nonsubscribers). Now, 13 percent of all unverified users were hitting the rate limit.

Users were outraged. If Musk wanted to stop scrapers, surely there were better ways than just cutting off access to the service for everyone on Twitter.

"Musk has destroyed Twitter's value & worth," wrote attorney Mark S. Zaid. "Hubris + no pushback–customer empathy–data = a great way to light billions on fire," wrote former Twitter product manager Esther Crawford, her loyalties finally reversed.

Musk retweeted a joke from a parody account: "the reason I set a 'View Limit' is because we are all Twitter addicts and need to go outside."

Aside from Musk, the one person who seemed genuinely excited about the changes was Evan Jones, a product manager on Twitter Blue. For months, he'd been sending executives updates regarding the anemic sign-up rates. Now, Blue subscriptions were skyrocketing. In May, Twitter had 535,000 Blue subscribers. At $8 per month this was about $4.2 million a month in subscription revenue. By early July, there were 829,391 subscribers—a jump of about $2.4 million in revenue, not accounting for App Store fees.

"Blue signups still cookin," he wrote on Slack above a screenshot of the sign-up dashboard.

Jones's team capitalized on the moment, rolling out a prompt to upsell users who'd hit the rate limit and encouraging them to subscribe to Twitter Blue. In July, this prompt drove 1.7 percent of the Blue subscriptions from accounts that were more than thirty days old and 17 percent of the Blue subscriptions from accounts that were less than thirty days old.

Yaccarino was notably absent from the conversation until July 4, when she shared a Twitter blog post addressing the rate limiting fiasco, perhaps deliberately burying the news on a national holiday.

"To ensure the authenticity of our user base we must take extreme measures to remove spam and bots from our platform," it read. "That's why we temporarily limited usage so we could detect and eliminate bots and other bad actors that are harming the platform. Any advance notice on these actions would have allowed bad actors to alter their behavior to evade detection." The company also claimed the "effects on advertising have been minimal."

If Yaccarino's role was to cover for Musk's antics, she was doing an excellent job. Twitter rolled back the limits shortly after her announcement.

On July 12, Musk debuted a generative AI company called xAI, which he promised would develop a language model that wouldn't be politically correct. "I think our AI can give answers that people may find controversial even though they are actually true," he said on Twitter Spaces.

Unlike the rival AI firms he was trying to block, Musk said xAI would likely train on Twitter's data.

"The goal of xAI is to understand the true nature of the universe," the company said grandly in its mission statement, echoing Musk's first, disastrous town hall at Twitter. "We will share more information over the next couple of weeks and months."

In November 2023, xAI launched a chatbot called Grok that lacked the guardrails of tools like ChatGPT. Musk hyped the release by posting a screenshot of the chatbot giving him a recipe for cocaine. The company didn't appear close to understanding the nature of the universe, but perhaps that's coming.

CHAPTER 68

"I'm Up for a Cage Match"

Musk's disastrous decision to roll out rate limits presented an opportunity for one of Twitter's major rivals just thirty miles down the road in Menlo Park.

Since March, Adam Mosseri, one of Mark Zuckerberg's top deputies at Meta and the head of Instagram, had been working on a Twitter competitor called Threads. The idea was to create an app that looked and felt like Twitter but ran on the decentralized protocol ActivityPub, allowing users to take their posts and followers with them if they left the platform.

Many tech leaders, including Mark Zuckerberg and Jack Dorsey, believed this was where social media was headed. "My view is that the more that there's interoperability between different services and the more content can flow, the better all the services can be," Zuckerberg said in an interview with *The Verge*'s Alex Heath. "And I guess I'm just confident enough that we can build the best one of the services, that I actually think that we'll benefit and we'll be able to build better quality products by making sure that we can have access to all of the different content from wherever anyone is creating it."

The Threads team was small and scrappy—just ten people, most of whom were designers and engineers. In July, Mosseri hired the security researcher Jane Manchun. Musk had tried to hire Wong in February 2023, then reached out again after she tweeted about him covering up the *W* on the TWITTER sign that hung over the company headquarters, so it said T ITTER instead of TWITTER.

"Want to work at Twitter?" Musk asked. "It's awesome." (It was unclear if he was referring to the company or the sign.)

Wong went to Meta instead.

Musk was annoyed that rival tech firms were benefiting from Twitter's struggles. Earlier that year, he downranked the corporate accounts belonging to TikTok, Snap, Meta, and Instagram, to stop their tweets from being recommended. (After months of falsely accusing Yoel Roth and Vijaya Gadde of "shadowbanning" institutions they didn't like, here was Musk, doing exactly that.) The move prompted a rare critique from Jack Dorsey, who told a user on the rival social platform Bluesky that Musk might not be the best person to run Twitter. "It all went south," he said vaguely. This, after Musk blocked people from sharing links to other social media sites on Twitter (which he had to reverse) and throttled links to the newsletter platform Substack, after it announced a Twitter-like feature called Notes. This final move prompted a strong rebuke from Matt Taibbi, who said he was finally leaving Twitter, although he has since continued tweeting.

Musk's tactics did not stop Twitter's traffic from dropping in the summer of 2023. In July, tweets per second, an internal metric used to track engagement on Twitter, was down by roughly fifteen hundred tweets, from six thousand tweets per second to just over forty-five hundred. Total active user seconds decreased globally around the world, most worryingly in the United States, United Kingdom, Latin America, and Canada.

Mosseri and Zuckerberg weighed the opportunity. If they launched in July, the app would be US-only to start. The European Union had recently rolled out a landmark piece of regulation known as the Digital Markets Act, or DMA, which put restrictions on Meta to stop it from mixing user data across its family of apps. "It's not just disclosures and consent. It's also verifying that there's no data leakage," Mosseri later told Casey Newton.

Meta decided to go for it, capitalizing on Twitter's self-inflicted death spiral. "It was either we wait on the EU, or delay the launch by many, many many months," Mosseri told Newton in a different conversation. "And I was worried that our window would close, because timing is important."

The bet instantly paid off. Within five days, Threads hit 100 million users, surpassing OpenAI's ChatGPT in its speed in hitting such a milestone (ChatGPT had taken two months to reach 100 million users after its launch).

Hours after Threads went live, Musk's lawyer Alex Spiro sent Zuckerberg a letter threatening to sue. "Twitter has serious concerns that Meta Platforms has engaged in systematic, willful, and unlawful misappropriation of Twitter's trade secrets and other intellectual property," he wrote.

Spiro accused Meta of hiring dozens of former Twitter employees to build a copycat app "with the specific intent that they use Twitter's trade secrets and other intellectual property in order to accelerate the development of Meta's competing app, in violation of both state and federal law as well as those employees' ongoing obligations to Twitter."

"[C]ompetition is fine, cheating is not," Musk tweeted grandly.

Meta responded with little more than a hand wave, as if brushing off a pesky fly. "No one on the Threads engineering team is a former Twitter employee—that's just not a thing," said Meta communications director Andy Stone.

Internally, Musk told employees the company needed to start shipping features faster. "And they need to be more robust," he said in a one-line email sent just before 3 a.m. Pacific time on July 11.

The launch cemented a bitter rivalry between Musk and Zuckerberg, who'd been feuding on and off for seven years. Musk, who'd been cash-poor while building Tesla and SpaceX, "brooded" about how easily Zuckerberg had made money in software, while Zuckerberg longed for the entrepreneurial esteem that Musk garnered in Silicon Valley, according to *The Wall Street Journal.*

Zuckerberg never had a reputation for being charming. He'd gotten better onstage and in interviews in the nearly two decades since he'd founded Facebook, but the average person would always associate the CEO with his portrayal in the film *The Social Network*, where Jesse Eisenberg played him as both aggressive and, at times, alien.

During the pandemic, Zuckerberg took up Brazilian jiujitsu and got impressively fit. After the CEO posted a photo of himself on Instagram, *The New York Times* described him "wearing a camouflage flak jacket while glistening faintly with sweat. His neck swells wider than his jaw. His shoulders are capped with muscle. His forearms bulge."

Now, Musk threw down the gauntlet. "I'm up for a cage match if he is lol," he tweeted.

The joke might've stopped there if Zuckerberg hadn't shared a screenshot of the post on his Instagram story, with a caption that read "send me location."

This was a version of Mark Zuckerberg that few people recognized. While Musk continued his childish antics on Twitter, saying "Zuck is a cuck" and suggesting the two men engage in a "a literal dick measuring contest 📏," Zuckerberg trolled him. When a Threads user pointed out that Twitter seemed to be blocking mentions of the app from trending on Twitter, Zuckerberg responded "Concerning 😂"—a signature Musk response. Zuckerberg's reputation as an out-of-touch nerd had softened after he took a page out of Musk's playbook: by shitposting.

Finally, after weeks of back and forth, wherein Musk shared seemingly fabricated updates about the cage match, offered to do a test run in Zuckerberg's backyard, and said he might need to have surgery, Zuckerberg gave up. "If Elon ever gets serious about a real date and official event, he knows how to reach me," the CEO posted on Threads. "Otherwise, time to move on. I'm going to focus on competing with people who take the sport seriously."

The initial buzz around Threads wore off within a couple months, and the app struggled to retain users. Mosseri resisted calls for Threads to lean into hard news. "Politics and hard news are inevitably going to show up on Threads—they have on Instagram as well to some extent—but we're not going to do anything to encourage those verticals," he wrote.

But his bet paid off in another way. "Around 2018, *The Information* reported Facebook has a 700-person PR team," Scott Galloway, a marketing professor at New York University's Stern School of Business, wrote on Threads. "Yet one person starched Meta's reputation clean, and he did it by accident." Finally, someone had taken the crown of least popular tech billionaire from Mark Zuckerberg.

CHAPTER 69

"We Shall Bid Adieu to the Twitter Brand, and Gradually, All the Birds"

On Monday, July 24, 2023, a bright orange crane rolled up to Twitter's San Francisco headquarters and got to work dismantling the prominent Twitter logo. Police arrived—the crane was blocking the street. Musk hadn't gotten the proper permits. The sign now read ER; the other letters lay scattered on the sidewalk below. Then the officers relented. This wasn't a police matter, they said, leaving the permits for the city to sort out.

The night before, Musk had let out a series of stream-of-consciousness tweets: Twitter was rebranding to X. "[S]oon we shall bid adieu to the twitter brand and, gradually, all the birds," he said.

Musk cemented the change in an email to employees the following morning. "We are indeed changing to X," Musk wrote just after 2:30 a.m. Pacific time on July 23. "And it is happening today. This is my last message from a Twitter email."

The move was met with a sigh of relief from some of Twitter's former employees. "I will be at peace with this," tweeted former VP of ad sales JP Maheu. "I have never been at peace with the name Twitter 2.0 as it implies it is still Twitter. The Twitter I knew and worked at for ten years no longer exist[s] since the acquisition."

"+1" wrote Lara Cohen, the former VP of partnerships and marketing.

"This is actually such a relief. The Twitter many of us knew and loved is gone."

Musk wasted no time changing Twitter's logo to a futuristic-looking white X set over a black background. In San Francisco, the conference rooms started getting renamed to things like "eXposure," "eXult," and "s3Xy." The company installed a strobing X sign on the roof. This also did not have a permit. The city issued a complaint. Inspectors visited the office on three different occasions in July but were denied access to the roof every time. A representative for the company claimed the sign was temporary.

In July 2023, Apple approved Twitter's name change in the App Store, bypassing a rule that barred developers from adopting single-character names. Twitter's transformation was complete.

Outside the company, people wondered aloud what the point of all this was. Why had Musk spent so much of his fortune on a brand he effectively planned to kill?

"I guess my question is, what was he paying for?" asked *Bloomberg* columnist Matt Levine. "Musk didn't want Twitter for its employees (whom he fired) or its code (which he trashes regularly) or its brand (which he abandoned) or its most dedicated users (whom he is working to drive away); he just wanted an entirely different Twitter-like service. Surely he could have built that for less than $44 billion? Mark Zuckerberg did!"

Finally, on July 31, Musk acquiesced to city officials and took down the X sign. The building had gone from TWITTER to T ITTER to X. And now, it was nothing.

After the name change went into effect, Linda Yaccarino, now CEO of X, tweeted, "X is the future state of unlimited interactivity—centered in audio, video, messaging, payments/banking—creating a global marketplace for ideas, goods, services, and opportunities. Powered by AI, X will connect us all in ways we're just beginning to imagine."

It was hard to discern what this meant. The company had launched video and messaging features, but the core of the platform remained unchanged. It was still a microblogging platform, albeit one with worse ads and more glitches than its predecessor. In the fall, Musk said he planned to put all of X behind a paywall, a move that seemed destined to scare off advertisers. One year after Musk acquired Twitter, he had managed to bring X to life—but in name only.

"How does it feel to see all the Twitter stuff gone?" I asked Doherty. He laughed. He nodded at the wall of his home office, where a massive bird logo glowed. As Musk was busy eradicating Twitter's signage from the office, Doherty had snagged the logo as a memento—and walked right out the door.

CHAPTER 70

"I Wish I Had Been More Worried"

Whispers that Linda Yaccarino was a CEO in name only had followed her since she took the job. In September 2023, she was scheduled to sit down with CNBC's Julia Boorstin at the Vox Media Code Conference in California to put the rumors to rest. Kara Swisher and Walt Mossberg had started the conference two decades prior; in 2022, Swisher announced she was stepping down. *The Verge*'s editor in chief, Nilay Patel, hosted the 2023 show, along with Boorstin and my colleague Casey Newton. Swisher attended as an interviewer.

On September 26, the day before Yaccarino was set to speak, Swisher threw a wrench into the plan. Yoel Roth, Twitter's former head of trust and safety, who'd been forced to go into hiding after Musk implied he was a pedophile, had agreed to be interviewed by Swisher at the event. It was a Kara Swisher special, one that would force Yaccarino to answer some uncomfortable questions during her talk. That is, if she still agreed to go onstage.

The morning of September 27, Swisher sent Yaccarino a text message to let her know about the change of plans. Later, Yaccarino claimed she'd received a forty-five-minute warning, implying she never received the message. To her credit, whenever she found out Roth was speaking, she didn't bail on the conference. The hosts gave her the option of speaking before Roth. Ultimately, she decided to go after. The audience would not be allowed to ask questions.

Later that day, Roth sat down for his interview with Swisher, expressing his concerns about the platform's ability to combat misinformation and hate speech. When Swisher asked him what advice he had for Linda Yaccarino, he said:

"Look at what your boss did to me. It happened to me. It happened after he sang my praises publicly. It happened after I didn't attack him. I didn't attack the company. And then he did that to me. I hope she is thinking about what those risks are and what she might face. If not for yourself, for your family, for your friends, for those that you love, be worried. You should be worried. I wish I had been more worried."

An hour later, Yaccarino walked onstage, looking rattled. Boorstin gave her the opportunity to respond to Roth's comments, which she said she'd be happy to do. "Yoel and I don't know each other," she began. "He doesn't know me, I don't know him. I work at X, he worked at Twitter. X is a new company, building a foundation based on free expression and freedom of speech. Twitter at the time was operating on a different set of rules, different philosophies and ideologies that were creeping down the road of censorship. It's a new day at X, and I'll leave it at that." A couple people clapped, which only underlined how many more did not.

Throughout the interview, Boorstin tried to pin Yaccarino down about how X was doing as a company, but Yaccarino didn't seem to know. She said X was close to breakeven and would turn a profit in 2024. But when asked about Musk's plan to put all of X behind a paywall, she seemed to have no idea what Boorstin was talking about. At one point, Boorstin asked the CEO about X's number of daily active users. Yaccarino responded that there were "probably 200, 250 [million], something like that."

The most famous image from the interview—the one that instantly went viral—was when Yaccarino held up her phone and inadvertently revealed X wasn't on her home screen. Instead, there were Facebook and Instagram, along with an app of the Holy Bible.

"The original super app," quipped *Verge* reporter Alex Heath.

CHAPTER 71

"The Stakes"

Early on Saturday, October 7, 2023, an unprecedented attack unfolded in the Middle East. Hamas militants bombarded Israel with rockets and breached the "Iron Wall" fence around Gaza, infiltrating the country and killing hundreds of civilians. Israel responded with airstrikes. By Monday morning, at least 900 Israelis had died, along with 560 Palestinians.

Almost instantly, X was awash in disinformation about the attacks. "Rather than being shown verified and fact-checked information, X users were presented with video game footage passed off as footage of a Hamas attack and images of firework celebrations in Algeria presented as Israeli strikes on Hamas," wrote David Gilbert in *Wired*. "There were faked pictures of soccer superstar Ronaldo holding the Palestinian flag, while a three-year-old video from the Syrian civil war repurposed to look like it was taken this weekend." It wasn't clear who was behind the disinformation campaigns—but researchers noted that the tactics were consistent with prior campaigns out of Iran.

Musk promoted two Twitter accounts known for spreading disinformation as credible news sources for the crisis. He deleted the tweet hours later, after it was viewed eleven million times.

As the violence unfolded, researchers manually tracked and debunked viral disinformation, a project made more difficult by the fact that Musk had cut off free API access and recently eliminated headlines from news articles. Musk had promoted Community Notes, a crowdsourced fact-checking tool, as his preferred news-vetting methodology. Two days after the attacks began, however, I searched for footage of the fireworks in Algeria being passed off as violence in Gaza, and seven of the top ten posts had no community note attached. Six were from Blue subscriber accounts. Some had thousands of views. Donald Trump Jr. posted a graphic video of the attack, only to get an incorrect note saying, "This is an old video and is not from Israel." The video was, in fact, real. Community Notes was no match for a rapidly unfolding global crisis.

X's collapse into disinformation and chaos was the culmination of many of Musk's decisions over the past year, including cutting off free API access for researchers, laying off trust and safety experts, and allowing anyone with $8 to buy a blue check mark.

Less than a week after the violence started, EU regulators opened an inquiry into X over the alleged spread of illegal content and disinformation, "in particular the spreading of terrorist and violent content and hate speech," which may have violated the Digital Services Act.

The need for a new platform to follow news in real time had never been more apparent. In the days that followed, Threads was more lively than it had been since the week after its debut. Yet Meta continued to resist calls to lean into news, wary of repeating past mistakes in amplifying misinformation. "We're not anti-news," Mosseri wrote. "News is clearly already on Threads. People can share news; people can follow accounts that share news. We're not going to get in the way of any either. But, we're also not going go to amplify news on the platform. To do so would be too risky given the maturity of the platform, the downsides of over-promising, and the stakes."

CHAPTER 72

"Zero Sum"

I met Randall Lin for the last time in the spring of 2023 at a steakhouse called Lolinda in the Mission District of San Francisco. As I approached the black door, I half expected Lin not to show. He'd landed a new job at OpenAI, the generative AI company that Musk had helped found and now saw as a major rival. In recent weeks, the machine-learning engineer had made it clear he wanted nothing to do with reporters. I'd asked him to meet me one last time to discuss his experience at Twitter.

I walked into the cacophonous dining room and was relieved to see Lin sitting at the table, noise-canceling headphones around his neck, beside his girlfriend, Alex. He seemed relaxed. When I asked him what he thought about Musk now, he paused, then recalled a story about a Tesla engineer who'd worked alongside him at Twitter.

The engineer said Musk was first and foremost a quality-control guy. He had spent a year optimizing Musk's commute, tweaking his Tesla to make microscopic changes until the CEO was satisfied.

Lin had once believed that Musk's passion for Twitter was an asset. If he could fix the platform for the user that cared about it most, he could improve the whole system, just like Tesla engineers tweaking Musk's car.

But now, with a few months of distance from the mercurial leader, Lin wasn't so sure. "Social media is zero sum," he said. "You can make his Tesla faster and it won't affect other people's driving experience. But with

Twitter, that's not true. We can make his experience less buggy and more fair, but we can't optimize his overall experience without it impacting everyone else."

Later that evening, I left the restaurant, waving goodbye to Lin and Alex as they walked down Mission Street. I thought again about his answer. It was ultralogical, demonstrating Lin's natural talent for engineering. He understood how Musk had negatively affected the experience for Twitter users. I wasn't so sure he'd grasped what Musk had done to him.

Musk might have been able to exterminate every bird logo in the office, but he couldn't kill tweeps' spirit. Throughout the layoffs, former employees came together in remarkable shows of solidarity and resilience. Hundreds of employees pooled money to hire lawyers to review the "hardcore" ultimatum. Those who landed new jobs hired former colleagues. Ex-employees fundraised for a coworker whose child had fallen ill, and for janitorial staff who'd been fired without severance. When Musk finally killed the bird logo, a creative strategy lead named Lee Owens gave away shirts with Twitter art. Eventually, a senior engineering manager named Menotti Minutillo coordinated with Owens and Shauna Wright, a content design lead, to sell the shirts in a fundraiser for the National Audubon Society, donating $11,500. "The culture didn't die with Twitter," a former executive tells me.

A year after the deal closed, Musk's legal fights with former Twitter employees were just beginning. By October 2023, X faced more than a dozen class-action suits, in addition to two thousand individual arbitration claims filed by employees still waiting on severance. X had stalled for nearly ten months, but finally agreed to participate in settlement negotiations. But at least one employee who was laid off during the many rounds of cuts after the acquisition was still collecting a paycheck as of publication, despite no longer having a job.

Parag Agrawal, Ned Segal, and Vijaya Gadde successfully sued X over $1.1 million in legal fees, which they incurred as a result of their work at the company.

"I have reviewed the amount in question, and although it is high and probably higher than most humans would like to pay, it's not unreasonable," said Delaware Court of Chancery judge Kathaleen St. J. McCormick in her decision—the same judge who presided over the Twitter-Musk trial twelve months earlier.

As of October 2023, the executives have not received the severance they are owed. Agrawal and Segal are still owed more than $100 million combined.

Since Yoel Roth had resigned, he'd arguably become the most recognizable face in the world of trust and safety. The accolade came at a serious cost. He had a bigger platform to share his message, but his safety, and the safety of his family, was at risk.

Roth understood that what had happened to him was not just a personal attack—it was a pressure campaign to stop content moderators from doing their jobs. Intentionally or not, Musk had taken the Trump playbook and applied it to the trust and safety field. He claimed it was about ethics, free speech, transparency. In reality, it was about silencing his perceived enemies. Musk didn't blame Jack Dorsey for Twitter's supposed missteps, though he'd been CEO at the time. He blamed Roth and Vijaya Gadde. It didn't seem like an accident.

Yao Yue, the infrastructure engineer who'd been fired in November, wanted to hold Elon Musk accountable for the way he'd treated her and her colleagues. In March, her lawyer, Shannon Liss-Riordan, filed charges with the National Labor Relations Board (NLRB) alleging that X fired Yue in retaliation for her Slack message and corresponding tweet in which she urged her colleagues not to resign.

"People often choose to accept illegal actions taken by people with more resources and better legal representation because they have no choice," Yue said. "When I did have a choice, I wanted to hold him accountable for myself and for people in similar situations."

In October 2023, Yue was vindicated, when the NLRB issued its first-ever complaint against X. The board alleged that X fired Yue as an act of retaliation and to discourage her colleagues from engaging in collective action.

The case is scheduled for a hearing in January 2024. If Yue wins, she could receive back pay for the months after she was fired. Already, it's a symbolic victory for tweeps. After news about the charge broke, Yue's phone blew up with messages from former colleagues. "Ex-tweeps are happy and encouraged by the development," she says modestly.

After leaving Twitter, Yue cofounded a company called IOP Systems with five former tweeps, building off the work they did to make infrastructure engineering more efficient. The cofounders set up their company so the power is split equally and no one has a majority share. Yue, who has two kids at home, gives people lots of latitude to decide where and when they work. "One lesson for me is that you have to defend your values," Yue says of her time at Twitter. "You can't just assume they'll survive." It's hard not to read her decisions as a direct response to working at a place that concentrated all its power in one man.

Oh, and one more thing. For those of you who read my cover story in *New York* in January 2023, Yao Yue is in fact Alicia. When I first gave her a pseudonym, I called her Emily, but she said it sounded "too good girl." She chose Alicia, after the character from *The Good Wife*.

Manu Cornet, who is part of an ongoing class-action suit against Twitter for unpaid severance, continued to document Twitter's unraveling even after he was fired. In October 2023, he published a book of Twitter cartoons, titled *Twittoons*. The tagline of the book reads: "One employee's cartoon chronicles Twitter's accelerated descent." He dedicated the book to Elon Musk, "without whose puerile mischief this book would have been very boring."

Cornet's beliefs about holding leaders accountable has changed. Before the takeover, the engineer had never considered joining a union. Why would he? He was a well-paid, in-demand programmer. Though the public sentiment toward unions has changed over the last decade, in tech there's still a strong belief that collective organizing can slow progress.

"I felt like a spoiled brat, given all the perks and stock options we enjoyed," Cornet later wrote. The takeover, the firings, the chaos—they changed his mind. "Collective action seems important now," he said.

One of Cornet's cartoons shows Musk talking to Mark Zuckerberg about the cage fight. "I got 100 million new users in 5 days," cartoon Zuckerberg says to Musk.

"I'm still much richer," Musk responds.

"Check out my pecs. I'd beat you in a cage fight," Zuckerberg says.

"But look at the size of my—" Musk begins.

"Elon . . ." Zuckerberg interrupts.

The final panel shows Musk and Zuckerberg standing on top of a skyscraper made of money. "Do you think people can hear us?" Zuckerberg asks.

"Maybe," Musk says. "But from down there, they are utterly unable to comprehend the subtlety of our rivalry."

In July, JP Doherty got his wish, finally landing a new job at Roblox. Musk had dangled stock options in front of Doherty in the spring, but in the end, even the promise of a payout wasn't worth it. Who could trust Musk, anyway? Tweeps were pretty sure the Spacebucks wouldn't amount to anything. For Doherty, it was finally time to leave.

"Greetings," his resignation email to Musk read. "Today will be my last day at X. I resign my position, effective 8/4/2023. Thank you for the opportunity to continue to serve the public conversation, and I wish you all continued success in your endeavors."

Musk replied with a single line: "Thanks JP."

In Slack, Doherty told his colleagues he was leaving. "After 11 and change years, it's time for me to move down the road," he said. "Thank you all. The people here have always made the difference." He signed it simply "over and out."

Doherty spent his last day walking around the office saying goodbye to his colleagues. It was the second time he'd done this in a decade. When he'd left Twitter for Apple in 2016, his colleagues had taken him out for drinks, and they'd spent the night walking around San Francisco, reminiscing fondly about their work.

Now, most of Doherty's colleagues simply looked deflated when he told them he was leaving. More than half told him that they, too, were trying to get out.

The first time Doherty left Twitter, he'd come back just four months later. This goodbye would be permanent.

On October 30, 2023, eighty-seven days after he resigned, Doherty was surprised to receive an email from X alerting him about new equity awards. Employees were getting new stock grants at $45 per share.

Doherty's award from early 2023 would vest in 2027, according to the paperwork, although he was leaning toward not accepting it, as he didn't want to have a financial tie to Musk's company.

Musk had promised this would happen seven months earlier, when he told employees the company was worth roughly $20 billion. Now, the valuation had dropped even further, to just $19 billion, or 43 percent of the purchase price.

In the disclosures, Musk reiterated his plan for X: to make it a town square, to defeat bots, to verify accounts. And finally, this line, which I found most revealing: "We believe that our success as an organization depends on our ability to implement these objectives. We cannot guarantee that we will be successful in implementing products or policies that further any of these objectives on a timely basis, or at all."

I flew to San Francisco to see Doherty in person shortly after he left Twitter. He picked me up from the Rockridge BART station, apologizing for the messy van, which his wife usually drove with the kids. In the back seat of the car sat a new beach wheelchair for his son, Rhys. Perhaps now Doherty would have time to help him use it.

Doherty said the week since he had left Twitter had been rough. He didn't like having time off. It made him antsy. Each morning, he'd wake up and check his phone, before remembering that he no longer had to worry about whether or not Twitter was still online.

"It's amazing to me that you can take so many years to build up the culture and goodwill, to the point that working at Twitter was really special, and then overnight it's, like, 'I could be fired at any moment. What the hell is today going to bring?'" he told me.

Doherty said he felt like a fraud for staying at Twitter as long as he did, even if, in his heart, he knew he had to put the well-being of his family before his moral outrage over Musk. "I was betraying my own principles by working there, because I thought that guy was a complete piece of shit, and I didn't want to work for him," he said. "But I didn't have a choice."

I asked Doherty if he still thought that Twitter would exist in a year. He paused for a moment.

"Have you ever been to the sandcastle contest in Alameda?" he asked, referring to a local event where dozens of people spent hours building intricate sandcastles on the Bay Area beach—a towering city, a sarcophagus, a car-size butterfly. I thought I could see where Doherty was going with the metaphor, that anything beautiful would eventually be washed away by the sea.

Doherty was actually going for something less romantic. "It's like an eight-year-old came and stomped on the castle," he said.

Conclusion

I'd thought once that if Twitter went away, I'd be grateful. The app was always noisy, distracting. As a woman, as a journalist, I'd never felt safe on the platform, particularly after Musk took over and the harassment got noticeably worse. In the summer of 2023, when Twitter became X and the app became so dysfunctional I stopped using it for anything outside of work, I was surprised to find that the void it left behind did not feel spacious. It felt sad. Previously, when something significant happened in the world, the news broke on Twitter first. People would react—with joy, with sadness, with jokes. Now the conversation was scattered, decentralized. It felt like I'd been shut out of a vast global conversation. Except, I hadn't. The conversation had simply become an echo chamber.

It's difficult to know how X is doing as a company, but we know the platform is losing users. In November 2022, Musk shared a screenshot on Twitter showing a high of 254.5 million monetizable daily active users. At the Code conference, Linda Yaccarino confirmed it was between 200 million and 250 million. It's not a huge decline, but it's not nothing. That same month, X's global traffic was down 14 percent year over year, and US traffic was down 19 percent, according to *TechCrunch*. Monthly ad revenue has continued to decline by at least 55 percent year over year every single month since Musk took over.

In October 2023, Yaccarino met with X's bankers, who'd loaned the company $13 billion one year earlier. Once again, she did her job, painting a positive picture of X's growth. "Not including the cost of servicing debt, the company already is cash flow positive," she claimed, according to *Bloomberg.*

Roughly a year after the acquisition was complete, Musk confirmed X was running a test to put the service behind a paywall, charging new users $1 to use the platform. The move is expected to slow growth and hurt advertising. Musk claimed the move would make it "1000x harder to manipulate the platform." Linda Yaccarino, the advertising expert, had no comment.

On September 6, 2023, Musk threatened to sue the Anti-Defamation League, a civil rights group that tracks hate speech on Twitter. The ADL's CEO, Jonathan Greenblatt, was initially supportive of Musk's acquisition. In October 2022, Greenblatt had excitedly mused, "to think what he can do with the public square." After the deal went through, though, the nonprofit reported a massive jump in hate speech on the platform. "Since the acquisition, The @ADL has been trying to kill this platform by falsely accusing it & me of being anti-Semitic," Musk said, seemingly referencing the study that prompted Twitter's trust and safety councilmembers to resign in December 2022.

Musk blamed the Jewish nonprofit for a continued 60 percent drop in US advertising revenue. "If this continues, we will have no choice but to file a defamation suit against, ironically, the 'Anti-Defamation' League," he tweeted. "If they lose the defamation suit, we will insist that they drop the 'anti' part of their name, since obviously . . . 😂"

It was hard to take Musk's claim, that the ADL was single-handedly responsible for all the disastrous business decisions Twitter had made over the past year, seriously. Musk's tweet sounded like he was broadly scapegoating the Jewish people. But it also didn't seem like an empty threat, given the lawsuit Musk filed against the Center for Countering Digital Hate one month prior. Later, Musk amplified an antisemitic conspiracy theory on X, prompting high-profile advertisers—including Apple, Disney, IBM, Paramount, and even Yaccarino's former company, NBCUniversal—to flee.

After Musk threatened to sue the ADL, I called a source at a prominent civil rights organization, who asked not to be named for fear their emails would become part of a future Twitter Files were they to ever speak out. The source said that they were still working with X, in a bizarre game of theater where they point out egregious examples of hate on the platform and X employees promise to take it down. Sometimes, the employees actually follow through. But the bar had moved, and the level of hate that's required for X to take action is higher than it was in Twitter's recent history. It was the outcome of Musk's original promise of realigning the platform to emphasize free expression over safety. Now there was more hate speech than ever, and the ADL was critical of X's decision to allow it.

Prior to Musk's takeover, the civil rights organization was often focused on high-follower hate accounts. These accounts warranted a lot of scrutiny, because they could direct their followers to harass people, and at times these harassment campaigns resulted in real-world harm. Elon Musk's account fit the description—an account that Twitter employees might have taken action against in another time.

When Musk initially sent his "extremely hardcore" email, he'd outlined what he believed it would take for the company to become lean, to build better products, to maximize its potential. To some, it sounded ambitious, exciting. Long hours would be worked and sacrifices would be made, all in the service of turning Twitter into the thing that Musk always believed it could be. Meanwhile, leaders across Silicon Valley looked at what was happening at Twitter, hoping to see if the labor movement that had flourished among the WFH class could be squashed, and if the progressive values of a newly self-actualized workforce could be brought back in line and made more productive.

But anyone seeking those answers discovered that the transition from

Twitter to X wrought something entirely different. Musk's intentions became clearer. In his mind, the company's success had nothing to do with people's work ethic or ability to think creatively. Instead, it was about placating the person at the top. Musk, after all, was the man with the vision. He was the one on the hero's journey.

It didn't matter how his behavior affected the workers around him; it certainly didn't matter if his voice was louder than any other user's.

Musk was the main character. It seemed like he really believed that.

ACKNOWLEDGMENTS

This book is a snapshot of the lives of Twitter employees during a pivotal moment in tech history. Thank you to the many, many tweeps who trusted me to tell this part of your story, but, in particular, JP Doherty, Yao Yue, and Yoel Roth, who all spent hours teaching me about the ins and outs of Twitter's culture.

The reporting in this book is the result of my four-year partnership with Casey Newton, who gave me my first job in journalism and continues to make me a better reporter, every day.

Kevin Nguyen, my editor at *The Verge*, was my collaborator on this book. Without his help, it would not have been written.

Thank you to my fact-checker, Becca Laurie, who pored over every line and spent hours on the phone with my sources. Any remaining errors are my own. James Herbert, who generously copy edited this book and suggested changes that made it so much better. My research assistant, Hannah Bassett, for her meticulous timeline. And Lora Kolodny and Walter Isaacson for generously sharing their reporting.

Merry Sun, my editor at Portfolio, for making this book so much better, and Laura Usselman, my agent, for making it happen in the first place. Thank you to the rest of the team at Portfolio, and Leila Sandlin, for their support throughout.

I'm indebted to Nilay Patel for taking a chance and hiring a newbie reporter, then pushing me to take on difficult stories. And to Alex Heath, who continues to break enviable scoops and gives me consistently sound advice.

Thank you most of all to my family and friends. My parents, for believing in me, and teaching me right from wrong. Eliana and Austin, for giving me a sense of belonging. Annie, Claire, and Allie, for inspiring me in all they do. And Andrew and Ava, who have my whole heart, and make all of this so worthwhile.

NOTES

EPIGRAPH

vii **Three sparks that set:** *The Divine Comedy of Dante Alighieri: Inferno*, trans. Allen Mandelbaum (New York: Bantam Dell, 2004); see also digitaldante.columbia.edu/dante/divine-comedy/inferno/inferno-6/.

vii **IF THE ZOO:** wint, Twitter post, May 22, 2012, 5:46 p.m., twitter.com/dril/status/205052027259195393.

INTRODUCTION

xiii **damp sock puppet:** Elon Musk, Twitter post, January 2022, 10:24 a.m., twitter.com/elonmusk/status/1486767109275328516.

xiv **the engineer, Yang:** Per source's request, I am using only his first name.

xiv **"This is a battle":** Elon Musk, Twitter post, November 28, 2022, 8:41 p.m., twitter.com/elonmusk/status/1597405399040217088.

xv **fled the platform:** Alexander Saeedy, Laura Cooper, and Alexa Corse, "Twitter's Revenue, Adjusted Earnings Fell About 40% in Month of December," *The Wall Street Journal*, March 3, 2023, wsj.com/articles/twitters-revenue-adjusted-earnings-fell-about-40-in-month-of-december-ee91f1eb.

xv **"why the absolute fuck":** kenny, Twitter post, February 13, 2023, 2:47 p.m., twitter.com/kenminkim/status/1625220231768707072.

xv **"My 'For You' page":** Tay, Twitter post, February 13, 2023, 5:45 p.m., twitter.com/TayInLA_/status/1625264934170009601.

xvi **"Is Twitter literally just":** John Sills, Twitter post, February 13, 2023, 2:44 p.m., twitter.com/johnjsills/status/1625219474147184640.

xvi ***Time* named Elon Musk:** Molly Ball, Jeffrey Kluger, and Alejandro de la Garza, "2021 Person of the Year: Elon Musk," *Time*, December 13, 2021, time.com/person-of-the-year-2021-elon-musk/.

xvi **Musk had recently asked his 62.8 million Twitter followers:** Walter Isaacson, *Elon Musk* (New York: Simon & Schuster, 2023).

xvi **57.9 percent voted yes:** Elon Musk, Twitter post, November 6, 2021, 3:17 p.m., twitter.com/elonmusk/status/1457064697782489088.

xvi **the deal closed in October 2022:** William Cohan, "Twitter's Debt Time Bomb & Linda Alarms," *Puck*, October 1, 2023, puck.news/twitters-debt-time-bomb-linda-alarms/.

xvi **he'd overpaid by roughly $19 billion:** Will Daniel, "Elon Musk's $44 Billion Twitter Purchase Is 'One of the Most Overpaid Tech Acquisitions in History,' Wedbush's Dan Ives Says. Twitter's Fair Value Is Only $25 Billion," *Fortune*, October 27, 2022, fortune.com/2022/10/27/elon-musk-twitter-purchase-most-overpaid-tech-history-dan-ives-wedbush/.

xvii **"Henry Ford of our time":** Ron Lampeas, "ADL Chief Favorably Compares Elon Musk to Antisemite Henry Ford, Then Thinks Again," *The Times of Israel*, October 8, 2022, timesofisrael.com/adl-chief-favorably-compares-elon-musk-to-antisemite-henry-ford-then-backtracks/.

xvii **"Going forward, to build a breakthrough Twitter 2.0":** Tom Warren, "Elon Musk Demands Twitter Employees Commit to 'Extremely Hardcore' Culture or Leave," *The Verge*, November 16, 2022, theverge.com/2022/11/16/23462026/elon-musk-twitter-email-hardcore-or-severance.

xvii **Anything that was "anti-meritocratic":** Ellise Shafer and McKinley Franklin, "Elon Musk Talks Twitter, Censorship and the 'Woke Mind Virus' on 'Real Time With Bill Maher,'" *Variety*, April 28, 2023, variety.com/2023/tv/news/elon-musk-bill-maher-twitter-censorship-1235584679/.

xvii **making the streaming service unwatchable:** Elon Musk, Twitter post, April 19, 2022, 10:10 p.m., twitter.com/elonmusk/status/1516600269899026432.

xvii **"I guess the times of complaining":** David Marcus, Twitter post, November 16, 2022, 2:57 p.m., twitter.com/davidmarcus/status/1592970137732538369.

xvii **He later credited Musk's example:** Steve Mollman, "Mark Zuckerberg Credits Elon Musk with Kicking Off the Tech Trend of Firing Middle Managers When Other CEOs Were 'a Little Shy,'" *Fortune*, June 9, 2023, fortune.com/2023/06/09/mark-zuckerberg-elon-musk-twitter-meta-layoffs-middle-managers/.

xvii **"Watching @elonmusk + Co":** John Herrman, "Why Elon Wants to Make Them Pay," *New York*, November 4, 2022, nymag.com/intelligencer/2022/11/why-elon-musk-wants-to-make-twitter-users-pay.html.

xviii **Six months after the deal closed:** Alex Hern, "Twitter's Value Down Two-Thirds Since Musk Takeover," *The Guardian*, May 31, 2023, theguardian.com/technology/2023/may/31/twitters-value-down-two-thirds-since-musk-takeover-says-investor.

PART 1: THE BIRD APP

3 **"Please ignore prior tweets":** Elon Musk, Twitter post, June 4, 2010, 2:31 p.m., twitter .com/elonmusk/status/15434727182.

3 **His father, Errol Musk:** Ashlee Vance, *Elon Musk: Tesla, SpaceX, and the Quest for a Fantastic Future* (New York: Ecco, 2017), 62.

4 **"I've got a million-dollar car":** "Watch a Young Elon Musk Get His First Supercar in 1999," CNN Vault, cnn.com/videos/business/2021/01/07/elon-musk-gets-his -mclaren-supercar-1999-vault-orig.cnn.

4 **"Watch this," Musk replied:** "PandoMonthly: Fireside Chat with Elon Musk," PandoDaily, July 17, 2012, YouTube, youtube.com/watch?v=uegOUmgKB4E.

4 **partially orchestrated by Thiel:** Vance, *Elon Musk: Tesla, SpaceX, and the Quest for a Fantastic Future.*

6 **Tesla's original founders, Martin Eberhard:** Jay Yarow, "Tesla Settles Lawsuit—Everyone Is a Founder," *Business Insider*, September 21, 2009, businessinsider .com/tesla-founders-end-bitter-legal-fight-2009-9.

6 **In SEC filings:** Form 8-K, Tesla Motors, Inc., United States Securities and Exchange Commission, November 5, 2013, sec.gov/Archives/edgar/data/1318605 /000119312513427630/d622890d8k.htm.

6 **For a guy worth $19.9 billion:** Profile, Elon Musk, *Forbes*, September 27, 2023, forbes.com/profile/elon-musk/?sh=5a741dc77999.

7 **a hundred fifty thousand individual shareholders:** Richard, Henderson, "Elon Musk Pits an Army of Tesla Fans Against Wall Street Analysts, Short Sellers," *Los Angeles Times*, February 11, 2020, latimes.com/business/story/2020-02-11/musk -tesla-fans.

7 **an army of Elon Musk fans:** "Matt Levine on Elon Musk: Chief Twit & 'Meme Lord,'" *On with Kara Swisher*, Spotify, October 2022, open.spotify.com/episode /0W0GzlE4j8iHBEdNo5HAKH.

7 **"I suspect that the Thai govt":** Elon Musk, Twitter post, July 4, 2018, 10:02 a.m., twitter.com/elonmusk/status/1014509856777293825.

7 **"Construction complete in about 8 hours":** Elon Musk, Twitter post, July 7, 2018, 2:39 p.m., twitter.com/elonmusk/status/1015666557458964480.

8 **"to make complimentary public statements":** Lora Kolodny, "Elon Musk Pressured Thai Officials to Say Nice Things About His Mini-Subs in the Midst of a Deadly Rescue Mission," CNBC, October 8, 2019, cnbc.com/2019/10/08/tesla-ceo-elon -musk-pressured-thai-officials-for-positive-pr.html.

8 **"We will make one":** Li Zhou, "Elon Musk and the Thai Cave Rescue: A Tale of Good Intentions and Bad Tweets," *Vox*, July 18, 2018, vox.com/2018/7/18/1757 6302/elon-musk-thai-cave-rescue-submarine.

8 **"Bet ya a signed dollar":** Alix Langone, "Elon Musk Calls a Diver That Rescued Thai Soccer Team a 'Pedo' on Twitter," *Time*, July 15, 2018, time.com/5339219 /elon-musk-diver-thai-soccer-team-pedo/.

8 **"I also did not literally mean":** Sara Randazzo and Tim Higgins "Elon Musk Takes the Stand in Lawsuit Accusing Him of Defamation Over 'Pedo' Tweet," *The Wall Street Journal,* December 3, 2019, wsj.com/articles/elon-musk-takes-the-stand-in-lawsuit-accusing-him-of-defamation-over-pedo-tweet-11575417397.

8 **Unsworth's lawyer on the case:** Anjali Huynh, "L. Lin Wood, Lawyer Who Tried to Overturn Trump's 2020 Loss, Gives Up License," *The New York Times,* July 6, 2023.

8 **permanently banned from Twitter:** Jack Brewster, "Lin Wood—Lawyer Closely Tied to Trump—Permanently Banned from Twitter After Claiming Capitol Siege Was 'Staged,'" Business and Human Rights Resource Center, January 7, 2021, business-humanrights.org/en/latest-news/lin-woodlawyer-closely-tied-to-trumppermanently-banned-from-twitter-after-claiming-capitol-siege-was-staged/.

8 **"Am considering taking Tesla":** Elon Musk, Twitter post, August 7, 2018, 12:48 p.m., twitter.com/elonmusk/status/1026872652290379776.

9 **he did not respect the SEC:** Lesley Stahl, "Tesla CEO Elon Musk: The 60 Minutes Interview," CBS News, December 9, 2018, cbsnews.com/news/tesla-ceo-elon-musk-the-2018-60-minutes-interview/.

9 **"The coronavirus panic":** Elon Musk, Twitter post, March 6, 2020, 3:42 p.m., twitter.com/elonmusk/status/1236029449042198528.

9 **shared transphobic beliefs ("Pronouns suck"):** Elon Musk, Twitter post, July 24, 2020, 11:42 p.m., twitter.com/elonmusk/status/1286869404874088448.

9 **"At least 50%":** Elon Musk, Twitter post, November 22, 2021, 12:20 a.m., twitter.com/elonmusk/status/1462652210739884035?lang=en.

10 **TWITTER IS RULING SXSW:** Michael Calore, *Wired,* March 9, 2007, wired.com/2007/03/twitter-is-ruling-sxsw.

11 **"Helicopter hovering above":** Mike Butcher, "Here's the Guy Who Unwittingly Live-Tweeted the Raid on Bin Laden," *TechCrunch,* May 2, 2011, techcrunch.com/2011/05/02/heres-the-guy-who-unwittingly-live-tweeted-the-raid-on-bin-laden-2.

12 **Amnesty International found:** Press release, "Crowdsourced Twitter Study Reveals Shocking Scale of Online Abuse Against Women," Amnesty International, December 18, 2018, amnesty.org/en/latest/press-release/2018/12/crowdsourced-twitter-study-reveals-shocking-scale-of-online-abuse-against-women.

12 **CEO Bob Iger said:** Peter Kafka, "Why Disney Didn't Buy Twitter," Vox, September 7, 2022, vox.com/recode/2022/9/7/23339402/bob-iger-disney-streaming-code.

13 **"We have seen abuse":** Jack Dorsey, "How Twitter Needs to Change," TED2019, April 2019, ted.com/talks/jack_dorsey_how_twitter_needs_to_change?language=en.

13 **Elliott Management acquired a 5 percent:** Lauren Feiner and Alex Sherman, "Twitter CEO Dorsey Keeps His Job After Company Strikes Investment Deal with Elliott Management, Silver Lake," CNBC, March 9, 2020, cnbc.com/2020/03/09/twitter-strikes-investment-deal-with-elliott-management-silver-lake-after-attempt-to-oust-ceo-jack-dorsey.html.

13 **Twitter employees rallied:** Ryan Mac, "Twitter Employees Show Support for Embattled CEO with #WeBackJack Hashtag," *BuzzFeed News*, March 2, 2020, buzzfeednews.com/article/ryanmac/webackjack-jack-dorsey-employee-support-elliott-activist.

14 **"I remember asking":** Casey Newton, "Going Down with the Censorships," *This American Life*, April 31, 2023, thisamericanlife.org/797/transcript.

15 **to Roth, it was a satisfying puzzle:** Newton, "Going Down with the Censorships," thisamericanlife.org/797/transcript.

15 **"facts about mail-in ballots":** Donald J. Trump, Twitter post, April 11, 2020, 8:29 p.m., twitter.com/realDonaldTrump/status/1249132374547464193.

15 **Twitter's new approach to labeling misinformation:** Yoel Roth and Nick Pickles, "Updating Our Approach to Misleading Information," Twitter Blog, May 11, 2020, blog.twitter.com/en_us/topics/product/2020/updating-our-approach-to-misleading-information.

15 **"How does a personality-free bag":** Yoel Roth, Twitter post, July 28, 2017, 1:55 a.m., twitter.com/yoyoel/status/890812999874691073.

16 **tried to roll out a new policy:** Deposition of Anika Collier Navaroli, Select Committee to Investigate the January 6th Attack on the US Capitol, US House of Representatives, September 1, 2022, january6th-benniethompson.house.gov/sites/democrats.january6th.house.gov/files/20220901_Anika%20Collier%20Navaroli.pdf.

16 **"Twitter saw President Trump's potential violent incitement":** "Social Media & the January 6th Attack on the US Capitol," draft, n.d., washingtonpost.com/documents/5bfed332-d350-47c0-8562-0137a4435c68.pdf?itid=lk_inline_manual_3.

23 **"Be there, will be wild!":** "Social Media & the January 6th Attack on the US Capitol."

23 **"lower the temperature on the platform":** "Social Media & the January 6th Attack on the US Capitol."

23 **hundreds more were injured:** Chris Cameron, "These Are the People Who Died in Connection with the Capitol Riot," *The New York Times*, January 5, 2022, nytimes.com/2022/01/05/us/politics/jan-6-capitol-deaths.html.

23 **The damages exceeded $2.7 million:** "24 Months Since the January 6 Attack on the Capitol," US Attorney's Office, US Department of Justice, updated January 4, 2023, justice.gov/usao-dc/24-months-january-6-attack-capitol.

24 **more than three hundred employees:** Kate Conger and Mike Isaac, "Inside Twitter's Decision to Cut Off Trump," *The New York Times*, January 16, 2021, nytimes.com/2021/01/16/technology/twitter-donald-trump-jack-dorsey.html.

24 **violation of Twitter's "Glorification of Violence" policy:** Company, Twitter Blog post, January 8, 2021, blog.twitter.com/en_us/topics/company/2020/Suspension.

24 **Twitter permanently suspended the former president:** "Permanent Suspension of @realDonaldTrump," Twitter Blog post, January 8, 2021, blog.twitter.com/en_us/topics/company/2020/suspension.

25 **Twitter's decision to ban Donald Trump:** "Social Media & the January 6th Attack on the US Capitol."

25 **Twitter actually amplified conservative voices:** Luca Belli, "Examining Algorithmic Amplification of Political Content on Twitter," X Blog, October 21, 2021, blog .twitter.com/en_us/topics/company/2021/rml-politicalcontent.

25 **"the free speech wing":** Amir Efrati, "Twitter CEO Costolo on Apple, Privacy, Free Speech and Google; Far from IPO," *The Wall Street Journal,* October 18, 2011, wsj.com/articles/BL-DGB-23367.

25 **Musk did not respect Trump:** Walter Isaacson, *Elon Musk* (New York: Simon & Schuster, 2023).

25 **Musk's eldest child, Jenna:** Isaacson, *Elon Musk.*

26 **"Unless the woke mind virus":** Isaacson, *Elon Musk.*

26 **"That's about the time":** "Exhibit H," Elon Musk text exhibits (Twitter v. Musk), September 28, 2022, documentcloud.org/documents/23112929-elon-musk-text -exhibits-twitter-v-musk.

27 **"I really like computer games":** Ashlee Vance, *Elon Musk: Tesla, SpaceX, and the Quest for a Fantastic Future.*

27 **"accelerant to creating X":** Dan Milmo and Amy Hawkins, "'The Everything App': Why Elon Musk Wants X to Be a WeChat for the West," *The Guardian,* July 29, 2023, theguardian.com/media/2023/jul/29/elon-musk-wechat-twitter-rebranding -everything-app-for-west#:~:text=Shortly%20before%20buying%20Twitter% 2C%20Musk,change%20is%20on%20its%20way.

28 **"someone you hate":** "Exhibit H," Elon Musk text exhibits (Twitter v. Musk).

28 **"Free speech is essential":** Elon Musk, Twitter post, March 25, 2022, 12:34 a.m, twitter.com/elonmusk/status/1507259709224632344.

29 **"Very much so!":** "Exhibit H," Elon Musk text exhibits (Twitter v. Musk).

29 **"My bullshit meter":** Cheyenne Ligon, "Elon Musk Reveals Talks with Sam Bankman-Fried: 'My Bulls**t Meter Was Redlining,'" Coindesk.com, updated May 9, 2023, coindesk.com/business/2022/11/12/elon-musk-reveals-talks-with-sam-bankman -fried-my-bullshit-meter-was-redlining.

29 **Bankman-Fried thought Musk was nuts:** Isaacson, *Elon Musk.*

32 ***New York Times* published an exposé:** Kate Conger, "Culture Changes and Conflict at Twitter," *The New York Times,* August 16, 2021, nytimes.com/2021/08/16 /technology/twitter-culture-change-conflict.html.

34 **But he filed a 13G:** Matt Levine, "Elon Musk Is Active Now," Bloomberg.com, April 6, 2022, bloomberg.com/opinion/articles/2022-04-06/elon-musk-is-active-now.

35 **That night, Agrawal pitched Musk:** Walter Isaacson, *Elon Musk.*

36 **"no, we didn't get the idea":** Twitter Comms, Twitter post, April 5, 2022, 5:31 p.m., twitter.com/TwitterComms/status/1511456430024364037.

36 **"Oh yeah . . . it had been":** "Exhibit H," Elon Musk text exhibits (Twitter v. Musk).

38 **Covid lockdowns on Twitter:** Elizabeth Dwoskin, "Elon Musk to Address Twitter Staff After Internal Outcry," *The Washington Post*, April 7, 2022, washingtonpost .com/technology/2022/04/07/musk-twitter-employee-outcry.

38 **retweeted a Hitler meme:** Elon Musk, Twitter post, January 30, 2022, 1:52 p.m., twitter.com/elonmusk/status/1487861173626101760.

38 **"I keep forgetting that you're still alive":** Elon Musk, Twitter post, November 14, 2021, 6:29 a.m., twitter.com/elonmusk/status/1459891238384115722.

39 **Employees were not pleased:** Elizabeth Dwoskin, "Elon Musk to Address Twitter Staff After Internal Outcry."

39 **"I'm extremely unnerved":** Dwoskin, "Elon Musk to Address Twitter Staff After Internal Outcry."

39 **"As expected," he said:** "Exhibit H," Elon Musk text exhibits (Twitter v. Musk).

41 **"I will provide advice":** "Exhibit H," Elon Musk text exhibits (Twitter v. Musk).

42 **"I'd love to learn more":** "Exhibit H," Elon Musk text exhibits (Twitter v. Musk).

42 **"You don't know how much":** Walter Isaacson, *Elon Musk*.

43 **"I don't feel anything":** Isaacson, *Elon Musk*.

43 **"Is Twitter dying?":** Elon Musk, Twitter post, April 9, 2022, 9:32 a.m., twitter.com /elonmusk/status/1512785529712123906.

43 **"What work did you get done":** "Exhibit H," Elon Musk text exhibits (Twitter v. Musk).

44 **"Btw, Parag is still":** "Exhibit H," Elon Musk text exhibits (Twitter v. Musk).

44 **"I came very close to dying":** Ashlee Vance, *Elon Musk: Tesla, SpaceX, and the Quest for a Fantastic Future*.

45 **"I am offering to buy":** Schdule 13D, Twitter, Inc., United States Securities and Exchange Commission, April 13, 2022, Exhibit B, sec.gov/Archives/edgar/data /1418091/ 000110465922045641/tm2212748d1_sc13da.htm.

46 **"I made an offer":** Elon Musk, Twitter post, April 14, 2022, 7:23 a.m., twitter.com /elonmusk/status/1514564966564651008.

46 **"No one believes this is the final price":** Akiko Fujiya, "Analyst on Musk's Twitter Offer: 'No Board in America Is Going to Take That Number,'" *Yahoo Finance*, April 14, 2022, finance.yahoo.com/news/no-board-in-america-is-going-to-take -that-number-analyst-on-musks-twitter-offer-190442334.html.

46 **Prince Alwaleed bin Talal:** Kateryna Kadabshy and Brandon Sapienza, "Billionaire Prince Alwaleed Rejects Elon Musk's Twitter Bid," Bloomberg.com, April 14, 2022, bloomberg.com/news/articles/2022-04-14/billionaire-prince-alwaleed -rejects-musk-s-twitter-bid?sref=CrGXSfHu.

46 **if Musk purchased 15 percent:** Lauren Feiner, "Twitter Board Adopts 'Poison Pill' after Musk's $43 Billion Bid to Buy Company," CNBC, April 15, 2022, cnbc .com/2022/04/15/twitter-board-adopts-poison-pill-after-musks-43-billion -offer-to-buy-company.html.

47 **"So, Elon, a few hours ago":** "Elon Musk Talks Twitter, Tesla and How His Brain Works—Live at TED2022," YouTube, April 14, 2022, youtube.com/watch?v=cd ZZpaB2kDM.

49 **"We are now collecting interest":** Lora Kolodny, "Start-up Investor Jason Calacanis Raising Millions of Dollars for Musk's Twitter Deal," CNBC, May 12, 2022, cnbc.com/2022/05/12/start-up-investor-jason-calacanis-raising-millions-for -twitter-stake.html#:~:text=CNBC%20viewed%20Calacanis's%20email %20to,deal%20will%20total%20about%20%2418%2C 000.

50 **revealed in court:** "Exhibit H," Elon Musk text exhibits (Twitter v. Musk).

50 **"Morgan Stanley and Jared":** "Exhibit H," Elon Musk text exhibits (Twitter v. Musk).

50 **By the end of April:** "S&P 500," Yahoo!Finance, finance.yahoo.com/quote/%5EG SPC/history.

50 **Twitter's banking advisers:** Michelle F. Davis and Liana Baker, "Twitter Takeover Was Brash and Fast, With Musk Calling the Shots," Bloomberg.com, April 26, 2022, bloomberg.com/news/articles/2022-04-26/twitter-takeover-was-brash-and -fast-with-musk-calling-the-shots?srnd=premium&sref=CrGXSfHu.

51 **The agreement stipulated:** Form 8-K, Tesla Motors, Inc., United States Securities and Exchange Commission, November 5, 2013, sec.gov/Archives/edgar/data /1418091/ 000119312522120474/d310843ddefa14a.htm.

51 **"Free speech is the bedrock":** "Elon Musk to Acquire Twitter," Securities and Exchange Commission, exhibit 99.1, sec.gov/Archives/edgar/data/1418091/ 000119 312522117720/d319190dex991.htm.

52 **"In the event of a layoff":** "Elon Musk and Twitter Were Sued by Six Ex-Employees in Delaware," CNBC, May 16, 2023, scribd.com/document/646717711/Elon-Musk -and-Twitter-were-sued-by-six-ex-employees-in-Delaware#.

53 **"if elon musk buys twitter":** Elizabeth Bruenig, Twitter post, April 14, 2022, 5:35 a.m., web.archive.org/web/20220414123940/https://twitter.com/ebruenig/status /1514583050964905984.

53 **"i was on tumblr":** lesbian mothman, Twitter post, April 14, 2022, 8:04 a.m., twitter .com/verysmallriver/status/1514575288319070214.

53 **"I started playing a game":** Dewayne Perkins, Twitter post, April 21, 2022, 9:20 p.m., twitter.com/DewaynePerkins/status/1517312377573679105.

53 **"Elon Musk was able":** Dividend Hero, Twitter post, April 25, 2022, 7:58 a.m., twit ter.com/HeroDividend/status/1518559997478813700.

53 **"Can someone just tell me":** Ned Miles, Twitter post, April 25, 2022, 3:02 p.m., twit ter.com/nedmiles/status/1518666741399928833.

53 **"I hope a weird guy":** Brooks Otterlake, Twitter post, April 25, 2022, 7:03 p.m., twit ter.com/i_zzzzzz/status/1518727355702808578.

53 **"If we want to raise":** Brittany Van Horne, Twitter post, April 26, 2022, 4:15 p.m., twitter.com/_brittanyv/status/1519047607607209992.

56 **"temporarily on hold":** Elon Musk, Twitter post, May 13, 2022, 5:44 a.m., twitter.com/elonmusk/status/1525049369552048129.

57 **to help fund the merger:** Twitter, Inc., Plaintiff, v. Elon R. Musk, X Holdings I, Inc., and X Holdings II, Inc., Defendants, Securities and Exchange Commission, exhibit 99.1, sec.gov/Archives/edgar/data/1418091/000119312522192993/d381911dex991.htm.

57 **Musk responded with:** Rachel Treisman, "Got a Question for Twitter's Press Team? The Answer Will Be a Poop Emoji," NPR, March 20, 2023, npr.org/2023/03/20/1164654551/twitter-poop-emoji-elon-musk.

61 **an eight-page letter:** Letter from Mike Ringler, Skadden, Arps, Slate, Meagher & Flom LLP to Twitter, Inc., July 8, 2022, sec.gov/Archives/edgar/data/1418091/ 000110465922078413/tm2220599d1_ex99-p.htm.

63 **"Musk had to try to conjure one of those":** Twitter, Inc., Plaintiff, v. Elon R. Musk, X Holdings I, Inc., Complaint, in the United States District Court for the State of Delaware, documentcloud.org/documents/22084462-twitter-v-elon-musk-complaint.

64 **"The longer the merger transaction":** Rebecca Kern, "Judge Sets Expedited Twitter v. Musk Trial for October," *Politico*, July 19, 2022, politico.com/news/2022/07/19/twitter-musk-lawsuit-expedited-trial- 00046533.

64 **the suit alleged:** Twitter, Inc., Plaintiff, v. Elon R. Musk, X Holdings I, Inc., Complaint, in the United States District Court for the State of Delaware, August 4, 2022, documentcloud.org/documents/22127591-musk-public-version-of-counterclaims-answer-w-cos.

64 **Segal emailed employees:** Kate Conger and Ryan Mac, "Twitter Tells Employees They Might Get Only Half Their Annual Bonus," *The New York Times*, August 19, 2022, nytimes.com/2022/08/19/technology/twitter-annual-bonus.html.

66 **After Mudge was fired:** Cara Lombardo, "Twitter Agreed to Pay Whistleblower Roughly $7 Million in June Settlement," *The Wall Street Journal*, September 8, 2022, wsj.com/articles/twitter-agreed-to-pay-whistleblower-7-million-in-june-settlement-11662661116.

66 **in its security apparatus:** Whistleblower Aid, "Twitter Whistleblower Disclosure," Whistleblower Aid, July 6, 2022, s3.documentcloud.org/documents/22186683/twitter-whistleblower-disclosure.pdf.

68 **"Even in cases":** Sheelah Kolhatkar, "Lina Khan's Battle to Rein in Big Tech," *The New Yorker*, November 29, 2021, newyorker.com/magazine/2021/12/06/lina-khans-battle-to-rein-in-big-tech.

70 **"Doing my best":** Elon Musk, Twitter post, July 7, 2022, 10:04 a.m., twitter.com/elonmusk/status/1545046146548019201.

71 **the CEO accepted his fate:** Anirben Sen and Tom Hals, "Musk Reverses Course, Again: He's Ready to Buy Twitter, Build 'X' App," Reuters, October 4, 2022, reu

ters.com/markets/europe/musk-said-go-ahead-with-5420-share-twitter-deal-bloomberg-reporter-2022-10-04.

71 **the number was closer to 25 percent:** "The Complete Elon Musk & BBC Interview (4/12/23) (Full Uncensored Footage)," Frosty Shadowhunter, April 12, 2023, youtube.com/watch?v=2SCgantWSn0&t=469s.

71 **"Some things are priceless":** "Elon Musk Reveals the Goal of the New Twitter," Fox News, April 18, 2023, youtube.com/watch?v=qbxStDmHFAE.

PART II: HELLSITE

75 **"Entering Twitter HQ—let that sink in!":** Elon Musk, Twitter post, October 26, 2022, 11:45 a.m., twitter.com/elonmusk/status/1585341984679469056.

76 **a trip that cost Twitter:** Joe Schneider, "Twitter Sued for Refusing to Pay for Two Private Jet Charters," Bloomberg.com, December 9, 2022, bloomberg.com/news/articles/2022-12-09/twitter-sued-after-refusing-to-pay-private-jet-charter-service-for-two-flights.

76 **planned to cut nearly 75 percent:** Elizabeth Dwoskin, Faiz Siddiqui, Gerrit De Vynck, and Jeremy B. Merrill, "Documents Detail Plans to Gut Twitter's Workforce," *The Washington Post*, October 20, 2022, washingtonpost.com/technology/2022/10/20/musk-twitter-acquisition-staff-cuts/.

76 **a coffee, microwaved please:** Walter Isaacson, *Elon Musk* (New York: Simon & Schuster, 2023).

76 **"I saw [Musk] as the guy":** Esther Crawford, Twitter post, July 26, 2023, 3:54 p.m., twitter.com/esthercrawford/status/1684291048682684416.

77 **grown up in a Christian cult:** Esther Crawford, Twitter post, April 16, 2022, 10:20 a.m., twitter.com/esthercrawford/status/1515379706010357760.

77 **she'd live-tweeted her contractions while giving birth:** Esther Crawford, Twitter post, May 7, 2009, 11:19 p.m., twitter.com/esthercrawford/status/1735284050.

77 **"sexual experiments and open relationships":** Emily Chang, "'Oh My God, This Is So F—ed Up': Inside Silicon Valley's Secretive, Orgiastic Dark Side," *Vanity Fair*, January 2, 2018, vanityfair.com/news/2018/01/brotopia-silicon-valley-secretive-orgiastic-inner-sanctum.

77 **"[M]y vibe check":** roon, Twitter post, October 26, 2022, 6:13 p.m., twitter.com/tszzl/status/1585394264351334400.

79 **"If we're not going anywhere":** "Not Going Anywhere," *MaNu* (Blog), April 27, 2022, ma.nu/blog/not-going-anywhere.

79 **design chief Dantley Davis:** Kate Conger, "Culture Change and Conflict at Twitter," *The New York Times*, August 16, 2021, nytimes.com/2021/08/16/technology/twitter-culture-change-conflict.html.

79 **hands Davis the iconic bird logo:** Manu Cornet, Twittoons, October 19, 2021, twittoons.com/4.

79 **"I took down the internal post":** "Bye Twitter," *MaNu* (Blog), November 1, 2022, ma.nu/blog/bye-twitter.

88 **"inviting bigots, liars, and grifters":** Jennifer Faull, "Will Advertisers Stick with a Twitter Run by Elon Musk?," *The Drum*, October 31, 2022, thedrum.com/news/2022/10/31/will-advertisers-stick-with-twitter-run-elon-musk.

88 **General Motors, a Tesla competitor:** Jamie L. La Reau, "GM Tweeting Has Largely Gone Silent as Musk Loses Millions in Ad Dollars," *Detroit Free Press*, November 30, 2022, freep.com/story/money/cars/general-motors/2022/11/30/gm-twitter-elon-musks-ads/69687219007/.

88 **General Mills, Audi, and Pfizer:** Yoel Roth, "I Was the Head of Trust and Safety at Twitter. This Is What Could Become of It," *The New York Times*, November 18, 2022, nytimes.com/2022/11/18/opinion/twitter-yoel-roth-elon-musk.html.

90 **Jack Dorsey also agreed to roll over:** Dan Primak, "Book Reveals Elon Musk's Secret Deal with Jack Dorsey," Axios, September 12, 2023, axios.com/2023/09/12/elon-musks-secret-deal-jack-dorsey-walter-isaacson-twitter.

91 **paid out for vested shares:** Lora Kolodny, "Twitter Tells Worried Employees Their Vested Shares Will Be Paid in Coming Days," CNBC, November 1, 2022, cnbc.com/2022/11/01/twitter-reassures-employees-vested-shares-will-be-paid-out-this-month.html.

92 **"You don't," he said:** Casey Newton, "Going Down with the Censorships," *This American Life*, April 31, 2023, thisamericanlife.org/797/transcript.

92 **"I showed him his own account":** Newton, "Going Down with the Censorships."

94 **Musk wanted users to be redirected:** Alex Heath, "Elon Musk Wastes No Time Changing Twitter," *The Verge*, October 30, 2022, theverge.com/2022/10/30/23430008/elon-musk-twitter-homepage-subscriptions-changes.

96 **"Ligma Johnson had it coming":** Elon Musk, Twitter post, October 28, 2022, 2:33 p.m., twitter.com/elonmusk/status/1586108809772089345.

97 **this violated Twitter's policy:** "The X Rules, Twitter Help Center, n.d., help.twitter.com/en/rules-and-policies/x-rules.

101 **"Who's critical?" they asked:** Zoë Schiffer, Casey Newton, and Alex Heath, "Extremely Hardcore," *New York*, January 17, 2023, nymag.com/intelligencer/article/elon-musk-twitter-takeover.html.

101 **"We envisioned a world":** Amir Shevat, "Developer Platforms Are All About Trust, and Twitter Lost It," *TechCrunch*, December 15, 2022, techcrunch.com/2022/12/15/developer-platforms-are-all-about-trust-and-twitter-lost-it.

102 **Tesla famously ran:** Lora Kolodny, "Elon Musk's Extreme Micromanagement Has Wasted Time and Money at Tesla, Insiders Say," CNBC, October 19, 2018, cnbc.com/2018/10/19/tesla-ceo-elon-musk-extreme-micro-manager.html.

102 **"you risk getting fired":** "Elon Musk and Twitter Were Sued by Six Ex-Employees in Delaware," CNBC, May 16, 2023, scribd.com/document/646717711/Elon-Musk-and-Twitter-were-sued-by-six-ex-employees-in-Delaware#.

103 drastically cut payroll costs: Elon Musk, Twitter post, November 4, 2022, 4:14 p.m., twitter.com/elonmusk/status/1588671155766194176.

105 a drunken fight: Kurtis Lee, "Elon Musk, in a Tweet, Shares Link from Site Known to Publish False News," *The New York Times*, October 30, 2022, nytimes.com/2022/10/30/business/musk-tweets-hillary-clinton-pelosi-husband.html.

106 "If I had a dollar": Elon Musk, Twitter post, October 31, 2022, 10:10 a.m., twitter.com/elonmusk/status/1587129795732770824.

108 "roughly half of US adults": Meltem Odabaş, "5 Facts About Twitter Lurkers," Pew Research Center, March 16, 2022, pewresearch.org/short-reads/2022/03/16/5-facts-about-twitter-lurkers.

109 "Twitter's current lords & peasants system": Elon Musk, Twitter post, November 1, 2022, 10:16 a.m., twitter.com/elonmusk/status/1587498907336118274.

109 "I wasn't really thinking": "'I Thought I'd Been Hacked. It Turned Out I'd Been Fired': Tales of a Twitter Engineer," Manu Cornet, *The Economist*, December 2, 2022, economist.com/1843/2022/12/02/i-thought-id-been-hacked-it-turned-out-id-been-fired-tales-of-a-twitter-engineer.

109 "$20 a month": Stephen King, Twitter post, October 31, 2022, 7:23 a.m., twitter.com/StephenKing/status/1587042605627490304.

109 "We need to pay the bills": Elon Musk, Twitter post, November 1, 2022, 10:16 a.m., twitter.com/elonmusk/status/1587312517679878144.

110 "When your team is pushing round": Esther Crawford, Twitter post, November 2, 2022, 12:34 a.m., https://twitter.com/esthercrawford/status/1587709705488830464.

110 "high-profile accounts on Twitter": Casey Newton and Zoe Schiffer, "Elon Only Trusts Elon," *Platformer*, December 5, 2022, platformer.news/p/elon-only-trusts-elon.

119 suit against X Corp. and Elon Musk : Courtney McMillian v. X Corp., f/k/a/ Twitter, Inc., Class Action Complaint, in the United States District Court, District of Northern California, July 12, 2023, si-interactive.s3.amazonaws.com/prod/planadviser-com/wp-content/uploads/2023/07/12150148/STAMPED-Twitter-Complaint-002.pdf.

120 agreed to represent them: KCAL, "Legal Actions Continue Against Twitter," KCAL News, December 5, 2022, cbsnews.com/losangeles/news/legal-actions-continue-against-twitter.

121 "Since we're all talking": Alex Haagaard, Twitter post, October 27, 2022, 5:46 p.m., twitter.com/alexhaagaard/status/1585749725768306690.

121 "Welcome to NEW TWITTER": Ben Rosen, Twitter post, October 27, 2022, 10:05 p.m., twitter.com/ben_rosen/status/1585814989683642368.

121 "maybe MAYBE ill stick around": Cullen Crawford, Twitter post, October 28, 2022, 11:58 a.m., twitter.com/HelloCullen/status/1586024569462149120.

121 "The scariest thing": 🦋 @simplylay, Twitter post, October 30, 2022, 7:16 p.m., twitter.com/simplylay/status/1586859593388335106.

121 **"People are like"**: Katie Notopoulos, Twitter post, November 1, 2022, 11:23 a.m., https://twitter.com/katienotopoulos/status/1587465303319003141.

121 **"for moral convenience"**: pathological supply avoidance, Twitter post, November 2, 2022, 10:06 a.m., twitter.com/griph/status/1587808323486863360.

121 **"'Sacks' is really"**: Alex Heath, Twitter post, November 3, 2022, 11:53 a.m., twitter .com/alexeheath/status/1588197607578312704.

121 **"The layoff email"**: Chris Bakke, Twitter post, November 4, 2022, 12:42 a.m., twit ter.com/ChrisJBakke/status/1588391239430246401.

122 **according to the Department of Justice:** United States of America v. Twitter, Inc., Opposition to X Corp.'s Motion for Protective Order & Relief from Consent Order, United States District Court Northern District of California, September 11, 2023, washing tonpost.com/documents/33a6f227-7259-4217-baf0-5d645b3c8b63.pdf.

124 **"leaked private information"**: Gergely Orosz, "The Scoop #32: Meta's Historic Layoffs," Pragmatic Engineer, November 17, 2022, newsletter.pragmaticengineer.com /p/the-scoop-32.

125 **"making such a risky change"**: Ryan Mac, Kate Conger, and Mike Isaac, "Twitter Is Said to Delay Changes to Check Mark Badges Until After Midterms," *The New York Times*, November 6, 2022, nytimes.com/2022/11/06/technology/twitter -verification-check-marks.html.

126 **announced Twitter Blue was already live:** Emma Roth, "Elon Musk's $7.99 Twitter Blue with Verification Is 'Coming Ssoon' on iOS," *The Verge*, November 5, 2022, theverge.com/2022/11/5/23442149/twitter-blue-checkmark-ios-update-elon -musk.

126 **"Power to the people"**: Casey Newton, "Elon Musk Has Discussed Putting All of Twitter Behind a Paywall," *The Verge*, November 7, 2022, theverge.com/2022 /11/7/23446262/elon-musk-twitter-paywall-possible.

126 **"The new Blue isn't live yet"**: Esther Crawford, Twitter post, November 5, 2022, 11:59 a.m., twitter.com/esthercrawford/status/1588969361976741888.

126 **"I just killed it"**: Elon Musk, Twitter post, November 9, 2022, 11:38 a.m., twitter .com/elonmusk/status/1590383366213611522.

126 **"I predict that over time"**: Jason Calacanis, Twitter post, November 3, 2022, 10:20 a.m., twitter.com/Jason/status/1588219571978190848.

127 **Calacanis needed to stop tweeting:** Kate Conger, Mike Isaac, Ryan Mac, "Two Weeks of Chaos: Inside Elon Musk's Takeover of Twitter," *The New York Times*, November 11, 2022, nytimes.com/2022/11/11/technology/elon-musk-twitter-takeover.html.

127 **"To be clear, Elon"**: Jason Calacanis, Twitter post, November 4, 2022, 2:02 p.m., twitter.com/Jason/status/1588637889952624641.

127 **"The blue checkmark simply"**: Valerie Bertinelli, Twitter post, November 5, 2022, 12:01 p.m., twitter.com/Wolfiesmom/status/1588969800675782656.

128 **tweeted right-wing pundit Benny Johnson:** Benny Johnson, Twitter post, November 6, 2022, 3:35 p.m., twitter.com/bennyjohnson/status/1589401140482883584.

129 Eli Lilly paused all its advertising: Drew Harwell, "A Fake Tweet Sparked Panic at Eli Lilly and May Have Cost Twitter Millions," *The Washington Post*, November 14, 2022, washingtonpost.com/technology/2022/11/14/twitter-fake-eli-lilly.

131 "Um, well, I'm a complaint hotline": "Elon Musk on Twitter Spaces Audio," YouTube, n.d., youtube.com/watch?v=r8ZhfmQHgKg.

132 fewer than sixty-one thousand users had subscribed: Matt Binder, "Surprise? Elon Musk's $8 Twitter Blue Hasn't made Very Much Money So Far," *Mashable*, November 11, 2022, mashable.com/article/twitter-blue-elon-musk-subscriber-numbers.

135 per a deposition with the FTC: United States of America v. Twitter, Inc., Opposition to X Corp.'s Motion for Protective Order & Relief from Consent Order, United States District Court Northern District of California, September 11, 2023, washingtonpost.com/documents/33a6f227-7259-4217-baf0-5d645b3c8b63.pdf.

135 "complete improvements in its data management": *Twitter, Inc.*, Opposition to X Corp.'s Motion for Protective Order & Relief from Consent Order, United States District Court Northern District of California, September 11, 2023.

136 "I don't watch Game of Thrones": Marianne Fogarty, Twitter post, November 7, 2022, 3:59 p.m., twitter.com/MarianneFogarty/status/1589724275510226944.

137 Yoel Roth wrote down: Newton, "Going Down with the Censorships."

137 "take arbitrary or unilateral": Newton, "Going Down with the Censorships."

138 "Behind Elon Musk": Newton, "Going Down with the Censorships."

141 "Elon Musk may be": TheCryptKeifer, Twitter post, November 3, 2022, 2:33 p.m., twitter.com/DannyVegito/status/1588237959236653057.

141 "being on Twitter right now": Lauren Dombrowski, Twitter post, november 6, 2022, 6:59 p.m., twitter.com/callmekitto/status/1589407148462964736.

141 "Every Elon post": John Frankensteiner, Twitter post, November 6, 2022, 7:45 p.m., twitter.com/JFrankensteiner/status/1589418718819647489.

141 "Everything happening on Twitter": Andrew Nadeau, Twitter post, November 7, 2022, 9:37 a.m., twitter.com/TheAndrewNadeau/status/1589628148446269442.

141 "every time the kind": Sarah Jeong, Twitter post, November 9, 2022, 8:12 p.m., twitter.com/sarahjeong/status/1590512775385976832.

143 Musk claimed impressions on hate speech: Elon Musk, Twitter post, December 2, 2022, 1:53 p.m., twitter.com/elonmusk/status/1598752139278532610.

143 an article that cited CCDH's research: Thomas Germain, "Hate Speech on Twitter Skyrocketed While Elon Said He Was Winning, New Research Shows," *Gizmodo*, December 2, 2022, gizmodo.com/elon-musk-twitter-hate-speech-data-n-word-rising-slurs-1849846553.

144 attempt to stifle free speech: Sheera Frenkel, "Twitter Sues Nonprofit That Tracks Hate Speech," *The New York Times*, July 31, 2023, nytimes.com/2023/07/31/technology/twitter-x-center-for-countering-digital-hate.html

145 "Btw, I'd like to apologize": Elon Musk, Twitter post, November 13, 2022, 1:00 p.m., twitter.com/elonmusk/status/1591853644944932865.

146 **she wrote in a quote tweet:** Sasha Solomon, Twitter post, November 13, 2022, 10:14 a.m., twitter.com/sachee/status/1591857120768843776.

147 **"After 12 amazing years":** Yao Yue, Twitter post, November 15, 2022, 6:18 a.m., twitter.com/thinkingfish/status/1592522358581264384.

148 **"i said it before and i'll say it again":** Sasha Solomon, Twitter post, November 14, 2022, 4:07 p.m., twitter.com/sachee/status/1592308273071681536.

150 **only about 25 percent of the software engineering organization:** Gergely Orosz, "The Scoop #32: Met's Historic Layoffs," Pragmatic Engineer, November 17, 2022, newsletter.pragmaticengineer.com/p/the-scoop-32.

154 **"The 'code reviews' were a clear pretext":** Wolfram Arnold et al., v. X Corp. f/k/a/ Twitter, Inc., Complaint, in the United States District Court for the State of Delaware, May 16, 2023, int.nyt.com/data/documenttools/twitter-employee-lawsuit-v/e5d27a60a7b7d51e/full.pdf.

154 **"The 'reviewers' lacked the context":** "Elon Musk and Twitter Were Sued by Six Ex-Employees in Delaware," CNBC, May 16, 2023, scribd.com/document/646717711/Elon-Musk-and-Twitter-were-sued-by-six-ex-employees-in-Delaware#.

155 **"rare earth metal":** Erin Griffith, "The Desperate Hunt for the A.I. Boom's Most Indispensable Prize," *The New York Times*, August 16, 2023, nytimes.com/2023/08/16/technology/ai-gpu-chips-shortage.html.

155 **Shares of Nvidia:** Derek Saul, "Nvidia Stock Hits All-Time High After 315% Surge—Easily Outpacing Its Peers," *Forbes*, August 22, 2023, forbes.com/sites/dereksaul/2023/08/22/nvidia-stock-hits-all-time-high-after-315-surge-easily-outpacing-its-peers.

156 **"app review teams had already begun":** Yoel Roth, "I Was the Head of Trust and Safety at Twitter. This Is What Could Become of It," *The New York Times*, November 18, 2022, nytimes.com/2022/11/18/opinion/twitter-yoel-roth-elon-musk.html.

157 **"Reinstate former President Trump":** Elon Musk, Twitter post, November 18, 2022, 7:47 p.m., twitter.com/elonmusk/status/1593767953706921985.

159 **reinstating thousands of accounts:** Clare Duffy, "The Mass Unbanning of Suspended Twitter Users Is Underway," CNN Business, December 8, 2022, cnn.com/2022/12/08/tech/twitter-unbanned-users-returning.

160 **"The reason Twitter sought":** "Elon Musk and Twitter Were Sued by Six Ex-Employees in Delaware," CNBC, May 16, 2023.

161 **the evening before Thanksgiving:** Zoë Schiffer, Twitter post, November 24, 2022, 8:35 a.m., twitter.com/ZoeSchiffer/status/1595773065962868737.

161 **four weeks of severance:** Alex Heath, Twitter post, November 24, 2022, 3:59 a.m., twitter.com/alexeheath/status/1595703616295567362.

162 **In the last two months of the year:** Jessica DiNapoli and Richa Naidu, "Focus: Here's What Twitter Lost in Advertising Revenue in Final Months of 2022," Reuters, January 20, 2023, reuters.com/technology/heres-what-twitter-lost-advertising-revenue-final-months-2022-2023-01-19/.

162 **Twitter's revenue in the fourth quarter of 2022:** Erin Woo, "Musk's Twitter Saw Revenue Drop 35% in Q4, Sharply Below Projections," *The Information*, January 18, 2023, theinformation.com/articles/musks-twitter-saw-revenue-drop-35-in-q4-sharply-below-projections.

162 **"until there is high confidence":** Elon Musk, Twitter post, November 21, 2022, 8:11 p.m., twitter.com/elonmusk/status/1594861031670820864.

164 **"Apple has mostly stopped advertising":** Elon Musk, Twitter post, November 28, 2022, 12:45 p.m., twitter.com/elonmusk/status/1597285572699074560.

164 **"threaten[ing] to withhold Twitter":** Elon Musk, Twitter post, November 28, 2022, 1:43 p.m., twitter.com/elonmusk/status/1597300125243944961.

164 **Phil Schiller, the Apple executive:** Sami Fathi, "Apple Executive Phil Schiller Deactivates Twitter Account," MacRumors, November 20, 2022, macrumors.com/2022/11/20/phil-schiller-deactivates-twitter-account.

165 **private data of Twitter Blue subscribers:** Walter Isaacson, *Elon Musk*.

166 **"This is the attitude":** Jay Peters, "Geohot Resigns from Twitter," *The Verge*, December 20, 2022, theverge.com/2022/12/20/23519922/george-hotz-geohot-twitter-internship-resigns.

166 **"Sure, let's talk":** Elon Musk, Twitter post, November 16, 2022, 3:23 p.m., twitter.com/elonmusk/status/1592976585858351105.

167 **The hacker believed:** "Transcript for George Hotz: Tiny Corp, Twitter, AI Safety, Self-Driving, GPT, AGI & God," *Lex Fridman Podcast*, podcast #387, June 29, 2023, lexfridman.com/george-hotz-3-transcript/#chapter12_working_at_twitter.

167 **a Slack message from Hotz:** Dave Beckett, Twitter post, May 23, 2023, 12:57 a.m., twitter.com/dajobe/status/1660872612837466112.

167 **"This will be awesome":** Elon Musk, Twitter post, December 2, 2022, 3:48 p.m., twitter.com/elonmusk/status/1598781280015073281.

168 **potentially explosive front-page story:** Emma-Jo Morris and Gabrielle Fonrouge, *New York Post*, October 14, 2020, nypost.com/2020/10/14/email-reveals-how-hunter-biden-introduced-ukrainian-biz-man-to-dad.

168 **The timing and placement:** Andrew Rice and Olivia Nuzzi, "The Sordid Saga of Hunter Biden's Laptop," *New York*, September 12, 2022, nymag.com/intelligencer/article/hunter-biden-laptop-investigation.html.

168 **WikiLeaks released a cache of emails:** Jeff Stein, "WikiLeaks Released a Cache of Emails," *Vox*, October 20, 2016, vox.com/policy-and-politics/2016/10/20/13308108/wikileaks-podesta-hillary-clinton.

169 **"The effect of such manipulations":** Jane Mayer, "How Russia Helped Swing the Election for Trump," *The New Yorker*, September 24, 2018, newyorker.com/magazine/2018/10/01/how-russia-helped-to-swing-the-election-for-trump.

169 **"obtained through hacking":** "Distribution of Hacked Material Policy, Twitter Rules and Policies," March 2019, web.archive.org/web/20190717143909/https://help.twitter.com/en/rules-and-policies/hacked-materials.

169 **Twitter took down the *New York Post*'s tweets:** Rice and Nuzzi, "The Sordid Saga of Hunter Biden's Laptop."

170 **"Twitter erred in this case":** "Statement of Yoel Roth, PhD," Hearing on "Protecting Speech from Government Interference and Social Media Bias, Part 1: Twitter's Role in Suppressing the Biden Laptop Story," House Committee on Oversight and Accountability, February 8, 2023, oversight.house.gov/wp-content/uploads/2023/02/Roth-House-Oversight-opening-statement-V4-Final.pdf.

170 **"Thread: THE TWITTER FILES":** Matt Taibbi, Twitter post, December 2, 2022, 6:34 p.m., twitter.com/mtaibbi/status/1598822959866683394.

171 **"The decision was made":** Matt Taibbi, Twitter post, December 2, 2022, 7:19 p.m., twitter.com/mtaibbi/status/1598834231794315265.

173 **Musk slashed Twitter's parental leave:** Kate Conger, Twitter post, May 7, 2022, 9:34 a.m., twitter.com/kateconger/status/1655249784574443520.

174 **"a desperate attempt to legitimize":** Joan Donovan, "Opinion: Why the 'Twitter Files' Are Falling Flat," *Politico*, December 15, 2022, politico.com/news/magazine/2022/12/15/twitter-files-falling-flat- 00073979.

175 **"Such listmakers are either":** Matt Taibbi, "Twitter Files: GEC, New Knowledge, and State-Sponsored Blacklists," Racket News, March 2, 2023, racket.news/p/twitter-files-gec-new-knowledge-and.

175 **"Tune in for Episode 2":** Elon Musk, Twitter post, December 2, 2022, 6:19 p.m., twitter.com/elonmusk/status/1598864377750306816.

175 **"Taibbi ftw":** Elon Musk, Twitter post, January 15, 2023, 6:59 p.m., twitter.com/elonmusk/status/1614819551694458880.

175 **"They're gonna keep bitching about it":** Dave Karpf, Twitter post, December 3, 2022, 10:39 p.m., twitter.com/davekarpf/status/1599171487402930176.

176 **did not go unnoticed by the FTC:** Interim Staff Report, "The Weaponization of the Federal Trade Commission: An Agency's Overreach to Harass Elon Musk's Twitter," Committee on the Judiciary and the Select Subcommittee on the Weaponization of the Federal Government, March 7, 2023, judiciary.house.gov/sites/evo-subsites/republicans-judiciary.house.gov/files/evo-media-document/Weaponization_Select_Subcommittee_Report_on_FTC_Harrassment_of_Twitter_3.7.2023.pdf.

177 **Chappelle got Covid after hanging:** Emily Kirkpatrick, "Dave Chappelle Tests Positive for COVID-19 After Hanging Out with Grimes, Elon Musk, and Joe Rogan," *Vanity Fair*, January 22, 2021, vanityfair.com/style/2021/01/dave-chappelle-tests-positive-coronavirus-joe-rogan-grimes-elon-musk.

178 **"Anyone recognize this person":** Elon Musk, Twitter post, December 14, 2022, 10:50 p.m., twitter.com/elonmusk/status/1603235998263123969.

179 **"locate his target":** Drew Harwell and Taylor Lorenz, "Musk Blamed a Twitter Account for an Alleged Stalker. Police See No Link," CBS News, December 18, 2022, washingtonpost.com/technology/2022/12/18/details-of-musk-stalking-incident/.

179 "to not banning the account": Elon Musk, Twitter post, November 6, 2022, 7:30 p.m., twitter.com/elonmusk/status/1589414958508691456.

179 "Remember Elon Musk's first Twitter Files": Matt Binder, Twitter post, December 24, 2022, 10:48 a.m., twitter.com/MattBinder/status/1606723471475605505.

180 "Same doxxing rules apply": Elon Musk, Twitter post, December 15, 2022, 6:12 p.m., twitter.com/elonmusk/status/1603573725978275841.

180 "You're a citizen": NowThis News, "Elon Musk Suspends Accounts, Accuses Journalists of Doxxing," YouTube, youtube.com/watch?v=nYezmbu_xRM.

181 fix a legacy bug: Elon Musk, Twitter post, December 16, 2022, 2:12 a.m., twitter.com/elonmusk/status/1603649264290123778.

181 "would never delete that account": Britney Nguyen, "Chamath Palihapitiya Says He's Faced Privacy Concerns over Jet-Tracking Like Elon Musk, and Would Consider Ditching Flying Private for Something 'More Anonymous,'" *Business Insider*, December 17, 2022, businessinsider.com/chamath-palihapitiya-on-private-jet-tracking-elon-musk-journalists-suspensions-2022-12.

181 "unsuspend accounts who doxxed": Elon Musk, Twitter post, December 15, 2022, 10:56 p.m., twitter.com/elonmusk/status/1603600001057185792.

184 "daily use of the n-word": CBS News Bay Area, "Watchdog Groups: Hate Speech Dramatically Surges on Twitter Following Elon Musk Takeover," CBS News, December 2, 2022, cbsnews.com/sanfrancisco/news/twitter-elon-musk-takeover-watchdog-groups-say-hate-speech-surges-dramatically.

184 "It is clear from research": Anne Collier, Twitter post, December 8, 2022, 11:25 a.m., twitter.com/annecollier/status/1600889250761027585/photo/1.

185 "You all belong in jail": Cernovich, Twitter post, December 9, 2022, 12:40 p.m. ET, twitter.com/Cernovich/status/1601270553650028546.

185 "It is a crime": Elon Musk, Twitter post, December 9, 2022, 12:59 p.m., twitter.com/elonmusk/status/1601275244710621184.

185 "Quit lying Ella": ruchowdh.bsky.social, Twitter post, December 10, 2022, 4:47 a.m., twitter.com/ruchowdh/status/1601513820035244037.

185 "and created more than 10": Joseph Menn, "Elon Musk Uses QAnon Tactic in Criticizing Former Twitter Safety Chief," *The Washington Post*, December 12, 2022, washingtonpost.com/technology/2022/12/12/musk-child-porn-qanon.

186 "Looks like Yoel is arguing": Elon Musk, Twitter post, December 10, 2022, 2:29 p.m., twitter.com/elonmusk/status/1601660414743687169.

187 the *Daily Mail* published an article: James Gordon, "Ex-Twitter Censor Yoel Roth and His Boyfriend Are Forced to FLEE Their $1.1m Home After His Thesis—Which Supports Letting Children Use Gay Hook-Up App Grindr—Is Shared by Elon Musk," *Daily Mail*, December 12, 2022, dailymail.co.uk/news/article-11531441/Ex-Twitter-censor-Yoel-Roth-boyfriend-forced-FLEE-t-1-1m-home-Elon-Musk-shared-thesis.html.

188 **cut $500 million from the budget:** "Musk Says Cost Cutting Averted $3 Billion Twitter Shortfall," Bloomberg.com, December 21, 2022, bloomberg.com/news/articles/2022-12-21/musk-says-cost-cutting-averted-3-billion-twitter-shortfall.

189 **the Goons succeeded:** Becky Peterson and Erin Woo, "Musk May Have Found a Hardcore Leader for Twitter," The Information, December 23, 2022, theinformation.com/articles/musk-may-have-found-a-hardcore-leader-for-twitter.

192 **"Well, we just won't pay those":** Wolfram Arnold et al., v. X Corp. f/k/a/ Twitter, Inc., Complaint, in the United States District Court for the State of Delaware, May 16, 2023, int.nyt.com/data/documenttools/twitter-employee-lawsuit-v/e5d27a60a7b7d51e/full.pdf.

192 **"Inside on the 10th floor":** Dave Beckett, Twitter post, June 5, 2023, 1:37 p.m., twitter.com/dajobe/status/1665774940597960706.

192 **her lawyers wrote:** Wolfram Arnold et al., v. X Corp. f/k/a/ Twitter, Inc., Complaint, in the United States District Court for the State of Delaware, May 16, 2023.

192 **according to a public baby registry:** "Nicole Hollander & Steven Davis's Registry," The Bump, n.d., registry.thebump.com/nicole-hollander-steven-davis-october-2022/54 000451.

194 **city inspectors came by the office:** "Complaint Data Sheet," City and County of San Francisco, December 6, 2022–August 22, 2023, dbiweb02.sfgov.org/dbipts/default.aspx?page=AddressComplaint&ComplaintNo=202299804.

200 **"If we lose one of those":** Donie O'Sullivan, Brian Fung, and Sean Lyngass, "Extreme California Heat Knocks Key Twitter Data Center Offline," CNN, September 12, 2022, cnn.com/2022/09/12/tech/twitter-data-center-california-heat-wave/index.html.

202 **"Even after I disconnected":** Elon Musk, Twitter post, December 24, 2022, 7:15 a.m., twitter.com/elonmusk/status/1606624671100997634.

PART III: MAIN CHARACTER

207 **"Each day on twitter":** maple cocaine, Twitter post, January 2, 2019, 10:20 p.m., twitter.com/maplecocaine/status/1080665226410889217.

208 **"The platonic ideal":** Michael Hobbes, Twitter post, August 27, 2023, 12:13 a.m., twitter.com/RottenInDenmark/status/1695650707159056440.

209 **the woman who made a pot of chili:** Emily Heil, "A Woman Made Chili for Neighbors. Outrage Ensued, Was She Wrong?," *The Washington Post*, November 18, 2022, washingtonpost.com/food/2022/11/18/chili-neighbors-twitter-etiquette/.

209 **the wife who loves sitting in her garden:** Daisey, Twitter post, October 21, 2022, twitter.com/lilplantmami/status/1583495587566977026.

209 **the young adult author:** Alexander McCoy, Twitter post, July 31, 2022, 9:40 p.m., twitter.com/AlexanderMcCoy4/status/1553918627581149185.

210 **The CEO had 124 million followers:** Elon Musk, Twitter profile, web.archive.org/web/20230102031613/twitter.com/elonmusk.

210 **the typical daily view count:** Faiz Siddiqui and Jeremy B. Merrill, "Elon Musk Reinvents Twitter for the Benefit of a Power User: Himself," *The Washington Post*, February 16, 2023, washingtonpost.com/technology/2023/02/16/elon-musk-twitter.

215 **tweeting about Dogecoin:** Elon Musk, Twitter post, February 12, 2023, 8:20 p.m., twitter.com/elonmusk/status/1624986889391407106.

215 **commenting on a video of a shirtless man:** Elon Musk, Twitter post, February 13, 2023, 2:07 p.m., twitter.com/elonmusk/status/1625255305868345345.

215 **"Vanity Unfair has fallen":** Elon Musk, Twitter post, February 13, 2023, 5:10 p.m., twitter.com/elonmusk/status/1625256224760696832.

216 **"MAJOR CHANGES TO THE ALGORITHM":** Zoe Schiffer and Casey Newton, "Yes, Elon Musk Created a Special System for Showing You All His Tweets First," *Platformer*, platformer.news/p/yes-elon-musk-created-a-special-system.

216 **"Please stay tuned":** Elon Musk, Twitter post, February 14, 2023, 12:11 a.m., https://twitter.com/elonmusk/status/1625407245218648065.

216 **"The 'source' of the bogus *Platformer*":** Elon Musk, Twitter post, February 17, 2023, 5:05 a.m., twitter.com/elonmusk/status/1626523188149764096.

217 **explosive report detailing alleged privacy violations:** Joseph Menn, "Ex-Twitter Engineer Tells FTC Privacy Violations Persist After Musk," *The Washington Post*, January 24, 2023, washingtonpost.com/technology/2023/01/24/whistleblower-twtter-ftc-settlement/.

217 **spying on behalf of Saudi Arabia:** Kevin Collier, "Former Twitter Employee Sentenced to More Than Three Years in Prison for Spying for Saudi Arabia," NBC News, December 14, 2022, nbcnews.com/tech/security/former-twitter-employee-sentenced-three-years-prison-spying-saudi-arab-rcna61384.

219 **Is everyone else's:** Zoë Schiffer, Twitter post, February 13, 2023, 5:33 p.m., twitter.com/ZoeSchiffer/status/1625261826979287040.

219 **"@ZoeSchiffer Yes":** Nicholas Brown, Twitter post, February 13, 2023, 10:03 p.m., twitter.com/News_By_Nick/status/1625329795528630274.

219 **"@ZoeSchiffer yes":** logan bartlett, Twitter post, February 13, 2023 https://twitter.com/loganbartlett/status/1625270333627420674.

219 **"@ZoeSchiffer Yes, WTF?":** David Weissman, Twitter post, February 13, 2023, 6:50 p.m., twitter.com/davidmweissman/status/1625281230500515841.

219 **"@ZoeSchiffer 100% but I was scared to":** Santiago Pombo, Twitter post, February 13, 2023, 3:47 p.m., twitter.com/SantiagoPombo/status/1625325742174400512

219 **"@ZoeSchiffer blocked him so no":** Khaver خاور, Twitter post, February 13, 2023, 6:48 p.m., twitter.com/thekarachikid/status/1625280694711734273

219 **"@ZoeSchiffer This is how Elon 'fixed' the algorithm":** Andrea Kuszewski, Twitter post, February 13, 2023, 7:28 p.m., twitter.com/AndreaKuszewski/status/1625290858970107904.

219 **"@ZoeSchiffer My man is paying":** anildash.com, Twitter post, February 13, 2023, 7:01 p.m., twitter.com/anildash/status/1625284010338385925.

219 **"@ZoeSchiffer 'For You' is":** Brian Ray, Twitter post, February 13, 2023, 12:44 a.m., twitter.com/brianrayguitar/status/1625370318503305216.

220 **"Twitter admits bias":** Dan Milmo, "Twitter Admits Bias in Algorithm for Rightwing Politicians and News Outlets," *The Guardian*, October 22, 2021, theguardian.com/technology/2021/oct/22/twitter-admits-bias-in-algorithm-for-rightwing-politicians-and-news-outlets.

221 **"good content that is free":** Emma Roth, "Elon Musk Says Bots with 'Good Content' Can Use Twitter's API for Free," *The Verge*, February 5, 2023, theverge.com/2023/2/5/23586577/elon-musk-bots-good-content-twitters-api-free.

221 **its new pricing:** Chris Stokel-Walker, "Twitter's $42,000-per-Month API Prices Out Nearly Everyone," *Wired*, March 10, 2023, wired.com/story/twitter-data-api-prices-out-nearly-everyone/.

221 **a number of "good" bots:** BART Alert, Twitter post, April 14, 2023, 8:19 p.m., twitter.com/SFBARTalert/status/1647031817889783812.

222 **"Sunlight is the best disinfectant":** Elon Musk, Twitter post, May 3, 2022, 3:35 p.m., twitter.com/elonmusk/status/1521574200183566338.

228 **"I made peace":** Esther Crawford, Twitter post, July 26, 2023, 3:54 p.m., twitter.com/esthercrawford/status/1684291048682684416.

228 **Musk replied, "It's approved":** Halli, Twitter post, March 6, 2023, 7:52 p.m., twitter.com/iamharaldur/status/1632906956229804032.

228 **Musk told his 129 million followers:** Elon Musk, Twitter profile, March 6, 2023, 11:47 p.m, twitter.com/elonmusk/status/1633011448459964417.

228 **"Not only is his work ethic":** Daniel Houghton, Twitter post, March 7, 2023, 9:42 a.m, twitter.com/danielhoughton/status/1633115945534214145.

229 **video call with Thorleifsson:** Elon Musk, Twitter post, March 7, 2023, 5:58 p.m., twitter.com/elonmusk/status/1633240643727138824.

231 **In contrast, Meta employees:** Mark Zuckerberg, "Mark Zuckerberg's Message to Employees," about.fb.com/news/2022/11/mark-zuckerberg-layoff-message-to-employees/

231 **two thousand former Twitter employees:** Fabien Ho Ching Ma v. Twitter, Inc., Petition to Compel Arbitration, United States District Court, Northern District of California, San Francisco Division, July 3, 2023, fingfx.thomsonreuters.com/gfx/legaldocs/akveqdzngvr/EMPLOYMENT_TWITTER_ARBITRATION_complaint.pdf.

233 **"Should be fixed now":** Elon Musk, Twitter post, February 17, 2023, 6:06 a.m., twitter.com/elonmusk/status/1626538656487059460.

233 **"the whole Sacramento shutdown":** Walter Isaacson, *Elon Musk* (New York: Simon & Schuster, 2023).

233 **the sixth major outage:** Casey Newton and Zoë Schiffer, "How a Single Engineer Brought Down Twitter on Monday," *Platformer*, March 6, 2023, platformer.news/p/how-a-single-engineer-brought-down.

233 **in all of 2022 Twitter experienced:** Ryan Mac, Mike Isaac, and Kate Conger, "'Sometimes Things Break': Twitter Outages Are on the Rise," *The New York Times*, February 29, 2023, nytimes.com/2023/02/28/technology/twitter-outages-elon-musk.html.

238 **"employees who purchased shares":** Zoë Schiffer, "The Secret List of Twitter VIPs Getting Boosted over Everyone Else," *Platformer*, March 27, 2023, platformer.news/p/the-secret-list-of-twitter-vips-getting.

238 **"Twitter, Inc. has been merged":** Laura Loomer, et al., v. Meta Platformers, Inc., Defendant Twitter, Inc.'s Corporate Disclosure Statement and Certification of Interested Entities or Persons, United States District Court, Northern District of California, San Francisco Division, April 4, 2023, storage.courtlistener.com/recap/gov.uscourts.cand.395023/gov.uscourts.cand.395023.123.0.pdf.

238 **the massive TWITTER sign:** Elon Musk, Twitter post, April 9, 2023, April 9, 2023, twitter.com/elonmusk/status/1645266104351178752.

239 **"many embarrassing issues":** Elon Musk, Twitter post, March 31, 2022, 2:46 p.m., twitter.com/elonmusk/status/1641874582473695246.

239 **the algorithm specifically labeled:** Jane Manchun Wong, Twitter post, March 31, 2023, 3:25 p.m., twitter.com/wongmjane/status/1641884551189512192.

239 **since he left the company:** Zoe Schiffer, "The Secret List of Twitter VIPs Getting Boosted Over Everyone Else," *Platformer*, March 27, 2023, platformer.news/p/the-secret-list-of-twitter-vips-getting.

240 **just 290,000 users:** Erin Woo, "Musk's Twitter Has Just 180,000 US Subscribers, Two Months After Launch," *The Information*, February 6, 2023, theinformation.com/articles/musks-twitter-has-just-180-000-u-s-subscribers-two-months-after-launch.

241 **"Welp guess my blue":** LeBron James, Twitter post, March 31, 2023, 12:16 p.m., twitter.com/KingJames/status/1641836984195743749.

241 **"Me joining you all":** Halle Berry, Twitter post, April 19, 2023, 10:02 p.m., twitter.com/halleberry/status/1648869732148301824.

242 **a #BlockTheBlueChecks campaign:** Mike Pearl, "Twitter Gives dril a Spite Checkmark," Mashable, April 22, 2023, mashable.com/article/twitter-verifies-dril-mashable-block-the-blue.

242 **a badge of dishonor:** Casey Newton, "Twitter's Badge of Dishonor," *Platformer*, April 24, 2023, platformer.news/p/twitters-badge-of-dishonor.

242 **"off my blue check":** Bobby Allyn, Twitter post, April 20, 2023, 8:26 p.m., twitter.com/BobbyAllyn/status/1649208070499872769.

242 **"Just before the purge":** Travis Brown, Twitter post, April 21, 2023, 11:24, a.m., twitter.com/travisbrown/status/1649433999914270722.

244 **"We want to enable users":** Alex Heath, "Twitter Is Making DMs Encrypted and Adding Video, Voice Chat, per Elon Musk," *The Verge*, November 21, 2022, theverge.com/2022/11/21/23472174/twitter-dms-encrypted-elon-musk-voice-video-calling.

244 **"those folks are badass":** Matthew Garrett, Twitter post, May 15, 2023, 6:15 p.m., twitter.com/mjg59/status/1658280059814432768.

245 **"Try it, but don't trust it yet":** Elon Musk, Twitter post, May 11, 2023, 1:03, a.m., twitter.com/elonmusk/status/1656570790039678976.

245 **"Twitter folks, seriously":** Gael Fashingbauer Cooper, "Twitter Now Offers Encrypted DMs, but Not Everyone Can Send Them," CNET, May 11, 2023. cnet.com/tech/services-and-software/twitter-now-offers-encrypted-dms-but-not-everyone-can-send-them/.

246 **the streaming giant canceled:** Alex Weprin, "Netflix Removed Titles in Turkey, Singapore Last Year Over Government Demands," *The Hollywood Reporter*, March 30, 2021, hollywoodreporter.com/business/business-news/netflix-removed-titles-in-turkey-singapore-last-year-over-government-demands-4158530/.

247 **Twitter received 550 requests:** "everything/2023/twitter-lumen/lumen-rest-of-world-project-data.csv," GitHub, n.d., github.com/row-engineering/everything/blob/main/2023/twitter-lumen/lumen-rest-of-world-project-data.csv.

247 ***Rest of World* reported:** Russel Brandom, "Twitter Is Complying with More Government Demands under Elon Musk," *Rest of World*, April 27, 2023, restofworld.org/2023/elon-musk-twitter-government-orders/.

247 **complied with a court order:** Chandni Shah, "Twitter Blocks Pakistan Govt's Account for Viewing in India—Notice," Reuters, March 29, 2023, reuters.com/world/asia-pacific/twitter-blocks-pakistan-govts-account-viewing-india-notice-2023-03-29.

247 **accounts that had been impacted:** "Twitter Succumbs to Erdoğan's Pressure, Silences Key Voices in Turkey on Election Eve," *Turkish Minute*, May 13, 2023, turkishminute.com/2023/05/13/twitter-succumbs-to-erdogan-pressure-silences-key-voices-in-turkey-on-election-eve.

248 **"The Turkish government":** Matt Yglesias, Twitter post, May 13, 2023, web.archive.org/web/20230513210815/https://twitter.com/elonmusk/status/1657422 40175425946.

248 **"The choice is":** Elon Musk, Twitter post, May 13, 2023, 12:27 p.m., twitter.com/elonmusk/status/1657422401754259461.

248 **"What Wikipedia did":** Jimmy Wales, Twitter post, May 13, 2023, 5:12 p.m., twitter.com/jimmy_wales/status/1657494022741426180.

249 **investigating Twitter's privacy practices:** Brian Fung, "FTC Says It's Conducting an Investigation into Twitter's Privacy Practices," CNN, March 8, 2023, cnn.com/2023/03/08/tech/ftc-twitter-privacy-investigation/index.html.

249 **"with privacy or information security":** "The Weaponization of the Federal Trade

Commission: An Agency's Overreach to Harass Elon Musk's Twitter," Committee on the Judiciary and the Select Subcommittee on the Weaponization of the Federal Government, March 7, 2023, judiciary.house.gov/sites/evo-subsites/republicans-judiciary.house.gov/files/evo-media-document/Weaponization_Select_Subcommittee_Report_on_FTC_Harrassment_of_Twitter_3.7.2023.pdf.

249 **"The Court should not permit the FTC":** United States of America v. Twitter, Inc., X Corp.'s Motion for Protective Order & Relief from Consent Order, United States District Court Northern District of California, August 17, 2023, washingtonpost.com/documents/184d9a3d-5aa5-4266-a501-31bb138a6e39.pdf.

250 **House Republicans jumped in:** Committee on the Judiciary and the Select Subcommittee on the Weaponization of the Federal Government, US House of Representatives, "The Weaponization of the Federal Trade Commission: An Agency's Overreach to Harass Elon Musk's Twitter," March 7, 2023, judiciary.house.gov/sites/evo-subsites/republicans-judiciary.house.gov/files/evo-media-document/Weaponization_Select_Subcommittee_Report_on_FTC_Harrassment_of_Twitter_3.7.2023.pdf.

250 **"was implemented so quickly":** United States of America v. Twitter, Inc., Opposition to X Corp.'s Motion for Protective Order & Relief from Consent Order, United States District Court Northern District of California, September 11, 2023.

251 **employees didn't have time:** *Twitter, Inc.,* Opposition to X Corp.'s Motion for Protective Order & Relief from Consent Order, United States District Court Northern District of California, September 11, 2023.

251 **"In fact, the relocated servers":** *Twitter, Inc.,* Opposition to X Corp.'s Motion for Protective Order & Relief from Consent Order, United States District Court Northern District of California, September 11, 2023.

251 **"easier to have more intimate":** Twitter Blog post, "Introducing Twitter Circle, a New Way to Tweet to a Smaller Crowd," August 30, 2022, blog.twitter.com/en_us/topics/product/2022/introducing-twitter-circle-new-way-tweet-smaller-crowd.

251 **those conversations were showing up:** Casey Newton, "Elon's War on Substack," *Platformer,* April 10, 2023, platformer.news/p/elons-war-on-substack.

251 **"Confirmed someone I'm not even following":** Theo—t3.gg, Twitter profile, twitter.com/t3dotgg/status/1644641624050520065.

253 **a condition of the agreement:** Martin Pengelly, "Tucker Carlson Claims in Book Fox News Firing Was Part of $787.5M Settlement," *The Guardian,* July 26, 2023, theguardian.com/books/2023/jul/26/tucker-carlson-fox-news-firing-condition-dominion-settlement.

253 **"Twitter is essentially following":** Charlie Warzel, "Twitter Is a Far-Right Social Network," *The Atlantic,* May 23, 2023, theatlantic.com/technology/archive/2023/05/elon-musk-ron-desantis-2024-twitter/674149/.

253 **an "uncancelable" free speech alternative:** Todd Spangler, "Parler Shut Down by New Owner: 'A Twitter Clone' for Conservatives Is Not a 'Viable Business,'" *Variety,* April 14, 2023, variety.com/2023/digital/news/parler-shut-down-new-owner-starboard-twitter-clone-conservatives-1235583709/.

253 **"Amazingly, as of tonight":** Tucker Carlson, Twitter post, May 9 2023, 4:42 p.m., twitter.com/TuckerCarlson/status/1656037032538390530.

254 **"On this platform":** Elon Musk, Twitter post, May 9 2023, 7:31 p.m., twitter.com/elonmusk/status/1656079504778092544.

254 **David Sacks, Musk's longtime collaborator:** Theodore Schliefer, "DeSantis-Sacks 24," Puck, May 23, 2023, puck.news/desantis-sacks-24.

255 **around 300,000 users listened in:** Ryan Mac and Tiffany Hsu, "DeSantis's Twitter Event Falls Short of the Reach of Past Livestreams," *The New York Times,* May 25, 2023, nytimes.com/2023/05/25/technology/ron-desantis-twitter-spaces-live-stream.html.

255 **campaign manager, Generra Peck:** Generra Peck, Twitter post, May 24, 2023, 8:08 p.m., twitter.com/GenerraPeck/status/1661524540416045058.

255 **including the internet and corporate media:** Josh Christenson and Jesse O'Neill, "DeSantis Camp Hawks 'Breaks Systems' Merch after Glitchy Presidential Announcement," *New York Post,* May 25, 2023, nypost.com/2023/05/25/desantis-camp-hawks-breaks-systems-merch-after-glitches/.

256 **"Tech giants like @youtube":** Jeremy Boreing, Twitter post, April 19, 2023, 3:39 p.m., twitter.com/JeremyDBoreing/status/1648773273583230982.

257 **"how a word as mundane":** "What Is a Woman," Matt's Movie Reviews, n.d., mattsmoviereviews.net/movie-critic-reviews/what-is-a-woman.html.

257 **"a film made to sermonize":** Jessie Gender, "Debunking Matt Walsh's 'What Is a Woman?,'" YouTube, youtube.com/watch?v=75bbNdlX2pA&ab_channel=JessieGender.

257 **In a lengthy tweet thread:** Jeremy Boreing, Twitter post, June 1, 2023, 8:59 a.m., twitter.com/JeremyDBoreing/status/1664255321630552065.

257 **Twitter responded to *The Daily Wire*'s pitch:** Jeremy Boreing, Twitter post, June 1, 2023, 8:59 a.m., twitter.com/JeremyDBoreing/status/1664255321630552065?lang=en.

257 **"deadnaming of transgender individuals":** GLAAD Press, "GLAAD Responds to Twitter's Roll-Back of Long-Standing LGBTQ Hate Speech Policy," *GLAAD,* April 18, 2023, glaad.org/releases/glaad-responds-twitters-roll-back-long-standing-lgbtq-hate-speech-policy.

258 **"My team was not told":** Ella Irwin, Twitter post, April 19, 2023, 1:00 p.m., twitter.com/ellagirwin/status/1648733338847281156.

258 **it had removed the line:** Jeremy Boreing, Twitter post, June 1, 2023, 5:59 a.m., twitter.com/JeremyDBoreing/status/1664255330711228417.

258 **"Of course, saying":** Jeremy Boreing, Twitter post, June 1, 2023, 8:59 a.m., twitter.com/JeremyDBoreing/status/1664255333882032128.

258 **"@elonmusk is not beholden":** Jeremy Boreing, Twitter post, June 1, 2023, 8:59 a.m., twitter.com/JeremyDBoreing/status/1664255337296191490.

258 **"Will Twitter make good":** Jeremy Boreing, Twitter post, June 1, 2023, 8:59 a.m., twitter.com/JeremyDBoreing/status/1664255339309531139.

258 **"This was a mistake":** Elon Musk, Twitter post, June 1, 2023, twitter.com/elonmusk /status/1664324213023424531.

259 **people who'd seen the tweet:** Daniel Chaitin, "'What Is A Woman?' Blows Past 170 Million Views on Twitter," *The Daily Wire,* June 5, 2023, dailywire.com/news/what -is-a-woman-blows-past-170-million-views-on-twitter.

259 **"Every parent should watch":** Elon Musk, Twitter post, June 2, 2023, 8:25 a.m., twitter.com/elonmusk/status/1664609193230204929.

259 **possible hateful conduct violation:** Greg Wehner, "Spotify CEO Question About Aliens Prompts Supreme Court Callback from Elon Musk: 'What Is a Woman?,'" Fox Business, May 29, 2023, foxbusiness.com/politics/spotify-ceo-question-aliens -prompts-supreme-court-callback-elon-musk-woman.

259 **"So one or two people noticed":** Ella Irwin, Twitter post, June 2, 2023, 7:56 p.m., https://twitter.com/ellagirwin/status/1664783096355381248.

259 **"Just kidding folks":** Ella Irwin, Twitter post, June 2, 2023, 8:04 p.m., twitter.com /ellagirwin/status/1664785100767395841.

260 **"Wake up tesla":** Phil Rosen, "This Investor Holds $74 Million in Tesla Stock. Here's Why He's Bullish on the Future of Elon Musk's EV Company," *Business Insider,* May 20, 2023, businessinsider.com/tesla-stock-elon-musk-ross-gerber -stock-twitter-bullish-investor-2023-5.

260 **Musk polled his Twitter followers:** Elon Musk, Twitter post, December 18, 2022, 6:20 p.m., twitter.com/elonmusk/status/1604617643973124097.

260 **One notable exception:** Robert F. Kennedy Jr., Twitter post, December 18, 2022, 9:23 p.m., twitter.com/RobertKennedyJr/status/16046652 00044646401.

261 **just 535,000 monthly Blue subscribers:** Angelique Chen, "Snap Reaches 1 mln Premium Subscribers in Bid for New Revenue," Reuters, August 15, 2022, reuters .com/technology/snap-reaches-1-mln-premium-subscribers-bid-new-revenue -2022-08-15/.

262 **The Starship rocket:** Marina Koren, "The Messy Reality of Elon Musk's Space City," *The Atlantic,* April 27, 2023, theatlantic.com/science/archive/2023/04/spacex -starship-explosion-dust-debris-texas/673881.

262 **"Excited to announce":** Hannah Murphy, "Why Linda Yaccarino Took on the Wildest Job in Silicon Valley," *Financial Times,* September 27, 2023, ft.com/con tent/18508c6f-c5f7-4e40-ad6f-ad46476d17bf.

262 **"cautiously optimistic" about Yaccarino:** Hannah Murphy and Daniel Thomas, "Twitter No Longer 'High Risk' After New Chief Hired, Says Top Ad Group," *Financial Times,* May 19, 2023, ft.com/content/1412d474-bd47-40be-a0a8-7f349b10e997.

262 **Twitter owed Google:** Ryan Mac, Tiffany Hsu, and Benjamin Mullin, "Twitter's New Chief Eases Into the Hot Seat," *The New York Times*, June 29, 2023, nytimes.com/2023/06/29/technology/twitter-ceo-linda-yaccarino.html.

265 **start sharing ad revenue:** Elon Musk, Twitter post, February 3, 2023, 11:21 a.m., twitter.com/elonmusk/status/1621544497388875777.

265 **Gigi Hadid deleted:** Alyssa Bailey, "Gigi Hadid Quit Twitter After Elon Musk's Takeover: 'It's Becoming More and More of a Cesspool of Hate,'" *Elle*, November 7, 2022, elle.com/culture/celebrities/a41885940/why-gigi-hadid-quit-twitter/.

265 **he was done "forever":** Wolfgang Ruth, "Every Celebrity Who Has Left the Bird App," *Vulture*, December 19, 2022, vulture.com/2022/12/celebrities-who-left-deactivated-twitter-elon-musk.html.

265 **Elton John stopped posting:** Elton John, Twitter post, December 9, 2022, 801 a.m., twitter.com/eltonofficial/status/1601200277276659712.

266 **Andrew Tate, a far-right influencer:** Beatrice Nolan, "Self-proclaimed Misogynist Andrew Tate Says He Was Paid $20,000 Under Elon Musk's Content-Creator Plan," *Business Insider*, July 14, 2023, businessinsider.com/andrew-tate-elon-musk-twitter-paid-content-creator-plan-2023-7.

266 **"hand-picked Twitter accounts":** Catturd, Twitter post, July 17, 2023, 11:17 a.m., twitter.com/catturd2/status/1680960011210481664.

268 **Data scraping was a known problem:** Ivan Mehta, "X Updates Its Terms to Ban Crawling and Scraping," *TechCrunch*, September 8, 2023, techcrunch.com/2023/09/08/x-updates-its-terms-to-ban-crawling-and-scraping.

269 **"The danger of training AI":** Elon Musk, Twitter post, December 16, 2022, 2:36 p.m., twitter.com/elonmusk/status/1603836383885332480.

269 **"our ability to crawl Twitter.com":** Jay Peters, "Tweets Aren't Showing Up in Google Results as Often Because of Changes at Twitter," *The Verge*, July 3, 2023, theverge.com/2023/7/3/23783153/google-twitter-tweets-changes-rate-limits.

270 **Rate limiting is a strategy:** "What Is Rate Limiting? Rate Limiting and Bots," Cloudflare, cloudflare.com/learning/bots/what-is-rate-limiting.

270 **"this sucks dude":** roon, Twitter post, July 1, 2023, 11:37 a.m., twitter.com/tszzl/status/1675212008025890817.

271 **"Musk has destroyed":** Mark S. Zaid, Twitter post, July 2, 2023, 5:44 p.m., twitter.com/MarkSZaidEsq/status/1675621440190595072.

271 **"Hubris + no pushback":** Esther Crawford, Twitter post, July 1, 2023, 4:90 p.m., twitter.com/esthercrawford/status/1675235246365880321.

271 **"the reason I set":** Elon Musk (Parody), Twitter post, July 1, 2023, 3:21 p.m., twitter.com/ElonMuskAOC/status/1675268446089773056.

272 **"effects on advertising":** "Updates on Twitter's Rate Limits," Business, Product Updates, n.d., business.twitter.com/en/blog/update-on-twitters-limited-usage.html.

272 **AI company called xAI:** Lora Kolodny, "Elon Musk Plans Tesla and Twitter Collaborations with xAI, His New Startup," CNBC, July 14, 2023, cnbc.com/2023/07/14/elon-musk-plans-tesla-twitter-collaborations-with-xai.html.

272 **"I think our AI":** Kolodny, "Elon Musk Plans Tesla and Twitter Collaborations with xAI, His New Startup."

272 **"The goal of xAI":** "Understand the Universe," Announcing XAI, July 12, 2023, x.ai.

273 **"My view is that":** Alex Heath, "Linda Yaccarino Set Up to Fail," *The Verge*, September 29, 2023, theverge.com/2023/9/29/23896248/linda-yaccarino-x-twitter-set-up-to-fail-command-line.

274 **"It all went south":** Ashley Capoot, "Jack Dorsey Criticizes Elon Musk's Leadership at Twitter: 'It All Went South,'" CNBC, April 29, 2023, cnbc.com/2023/04/29/jack-dorsey-criticizes-elon-musks-leadership-at-twitter-it-all-went-south.html.

274 **Musk blocked people from sharing links:** Jay Peters, "X Seemed to Throttle Some Competitors and News Sites for More Than a Week," *The Verge*, August 15, 2023, theverge.com/2023/8/15/23833314/x-twitter-throttling-traffic-competitors-news-sites-elon-musk.

274 **he has since continued tweeting:** Mitchell Clark, "One of Elon's Handpicked 'Twitter Files' Writers Quits Twitter over Its Substack Restrictions," *The Verge*, July 6, 2023, theverge.com/2023/4/7/23674705/twitter-files-elon-musk-substack-matt-taibbi.

274 **"It's not just disclosures":** Casey Newton, host, Hard Fork, "Special Episode: Meta's Twitter Rival Arrives, with Adam Mosseri," *The New York Times*, July 6, 2023, nytimes.com/2023/07/06/podcasts/special-episode-metas-twitter-rival-arrives-with-adam-mosseri.html.

275 **Meta decided to go for it:** Casey Newton, "Meta Unspools Threads," *Platformer*, July 5, 2023, platformer.news/p/meta-unspools-threads.

275 **"with the specific intent that they use Twitter's trade secrets":** Aaron Katersky, "Twitter Sends Meta Cease-and-Desist Letter Over New Threads App: Sources," ABC News, July 7, 2023, abcnews.go.com/Business/twitter-sends-meta-cease-desist-letter-new-threads.

275 **"[C]ompetition is fine":** Elon Musk, Twitter post, July 6, 2023, 3:51 p.m., twitter.com/elonmusk/status/1677042708756439041.

275 **"No one on the Threads engineering team":** Max Tani, "Twitter Is Threatening to Sue Meta over Threads," Semafor, July 6, 2023, semafor.com/article/07/06/2023/twitter-is-threatening-to-sue-meta-over-threads.

275 **The launch cemented a bitter rivalry:** Tim Higgins and Deepa Seetharaman, "Behind the Musk-Zuckerberg 'Cage Match' Is a Yearslong Billionaire Feud," *The Wall Street Journal*, June 23, 2023, wsj.com/articles/elon-musk-mark-zuckerberg-cage-fight-b71fa4ff.

276 **posted a photo of himself:** Zuck, Instagram post, May 29, 2023, instagram.com/p/CslpltwPx1a.

276 **"wearing a camouflage flak jacket":** Joseph Bernstein, "Mark Zuckerberg Would Like You to Know About His Workouts," *The New York Times*, June 2, 2023, nytimes.com/2023/06/02/style/mark-zuckerberg-workout.html.

276 **"I'm up for a cage match":** Elon Musk, Twitter post, June 20, 2023, 11:50 p.m., twitter.com/elonmusk/status/1671364992665264131.

276 **"send me location":** Christopher Brito, "Mark Zuckerberg Agrees to Fight Elon Musk in Cage Match: "Send Me Location," CBS News, June 22, 2023, cbsnews.com/news/mark-zuckerberg-elon-musk-fight.

276 **"Zuck is a cuck":** Elon Musk, Twitter post, July 9, 2022, 10:45 a.m., twitter.com/elonmusk/status/1678098028849143809.

276 **"a literal dick measuring contest":** Elon Musk, Twitter post, July 9, 2022, 7:01 p.m., twitter.com/elonmusk/status/1678222776908275712.

276 **"Concerning":** Mark Zuckerburg, Threads post, July 9, 2023, threads.net/@zuck/post/Cudy3UXOiKT.

276 **"If Elon ever gets serious":** Julianne McShane, "Zuckerberg Dismisses Musk for Avoiding Cage Fight: 'It's time to move on'" NBC News, April 13, 2023, cnbc.com/2023/08/13/zuckerberg-dismisses-musk-for-avoiding-cage-fight-its-time-to-move-on.html.

277 **"encourage those verticals":** mosseri, Threads post, July 7, 2023, 10:12 a.m., threads.net/@mosseri/post/CuZ3LjhNl0m.

278 **"[S]oon we shall bid adieu":** Elon Musk, Twitter post, July 23, 2023, twitter.com/elonmusk/status/1682964919325724673.

278 **"I have never been at peace":** JP Maheu, Twitter post, July 23, 2023, 12:19 p.m., twitter.com/jpmaheu/status/1683149745362030592.

279 **"eXposure," "eXult," and "s3Xy":** Ryan Ma and Tiffany Hsu, "From Twitter to X: Elon Musk Begins Erasing an Iconic Internet Brand," *The New York Times*, July 24, 2023, nytimes.com/2023/07/24/technology/twitter-x-elon-musk.html.

279 **The city issued:** "Complaint Data Sheet," City and County of San Francisco, July 28–August 7, 2023, dbiweb02.sfgov.org/dbipts/Default2.aspx?page=AddressComplaint&ComplaintNo=202311248.

279 **Apple approved Twitter's name change:** Ivan Mehta, "Apple Greenlights, Twitter's Rebrand to X," *TechCrunch*, July 31, 2023, techcrunch.com/2023/07/31/apple-greenlights-twitter-apps-rebrand-to-x.

279 **"what was he paying for?":** Matt Levine, "The APEs Can Be Saved," Bloomberg.com, July 24, 2023, bloomberg.com/opinion/articles/2023-07-24/the-apes-can-be-saved.

279 **"X is the future state":** Linda Yaccarino, Twitter post, July 23, 2023, 1:33 p.m., twitter.com/lindayaX/status/1683213895463215104.

282 **"I wish I had been more worried":** *The Verge*, "Yoel Roth Warns New X CEO about Elon and Company Status [FULL INTERVIEW]," YouTube, youtube.com/watch?v=M9XoUUYeZD8.

282 **"It's a new day at X":** Chas Danner, "What X CEO Linda Yaccarino Claimed at the Code Conference," *New York*, updated September 28, 2023, nymag.com/intelligencer/2023/09/what-x-ceo-linda-yaccarino-claimed-at-the-code-conference.html.

282 **"the original super app":** Alex Heath, "Linda Yaccarino Set Up to Fail," *The Verge*, September 29, 2023, theverge.com/2023/9/29/23896248/linda-yaccarino-x-twitter-set-up-to-fail-command-line.

283 **By Monday morning:** Isabel Kershner, Aaron Boxerman, and Hiba Yazbek, "Israel Orders 'Complete Siege' of Gaza and Hamas Threatens to Kill Hostages," *The New York Times*, October 9, 2023, nytimes.com/2023/10/09/world/middleeast/israel-gaza-siege-hamas.html.

283 **X was awash in disinformation:** David Gilbert, "The Israel-Hamas War Is Drowning X in Disinformation," *Wired*, October 9, 2023, wired.com/story/x-israel-hamas-war-disinformation/.

283 **prior campaigns out of Iran:** Moderated Content, "MC Weekly Update 10/9: Social Media During War," *Stanford Law School Podcasts*, October 10, 2023, law.stanford.edu/podcasts/mc-weekly-update-10-9-social-media-during-war/.

283 **after it was viewed eleven million times:** Joseph Menn, "As False War Information Spreads on X, Musk Promotes Unvetted Accounts," *The Washington Post*, October 8, 2023, washingtonpost.com/technology/2023/10/08/israel-hamas-disinfo-musk-twitter-x/.

284 **his preferred news-vetting methodology:** Elon Musk, Twitter post, February 21, 2023, 3:30 p.m., twitter.com/elonmusk/status/1628175431315644419.

284 **"This is an old video and is not from Israel":** David Gilbert "A Graphic Hamas Video Donald Trump Jr. Shared on X Is Actually Real, Research Confirms," *Wired*, October 11, 2023, wired.com/story/x-community-notes-failures/.

284 **rapidly unfolding global crisis:** Ben Goggin "Inside X's Community Notes, Fact-Checks on Known Misinformation Are Delayed for Days," NBC News, October 10, 2023, nbcnews.com/tech/misinformation/elon-musk-x-fact-check-israel-misinformation-rcna119658.

284 **EU regulators opened an inquiry:** European Commision, "The Commission Sends Request for Information to X under the Digital Services Act," October 12, 2023, ec.europa.eu/commission/presscorner/detail/en/IP_23_4953.

284 **"We're not anti-news":** Sarah Perez, "Instagram Head Says Threads Is 'Not Going to Amplify News on the Platform,'" *TechCrunch*, October 11, 2023, techcrunch.com/2023/10/11/instagram-head-says-threads-is-not-going-to-amplify-news-on-the-platform/.

286 **waiting on severance:** Matt Binder, "Elon Musk's X Finally Agrees to Settlement Talks with Unpaid Laid-Off Twitter Employees," Mashable, September 14, 2023, mashable.com/article/twitter-x-elon-musk-laid-off-employees-severance-negotiations.

287 **Ned Segal and Vijaya Gadde successfully sued:** Jef Feeley, "Twitter Execs Win $1.1 Million in Legal Fees From Musk's X," *Bloomberg*, bloomberg.com/news/articles/2023-10-03/ex-twitter-executives-win-1-1-million-legal-fees-from-musk-s-x.

287 **more than $100 million combined:** Theo Francis, "Twitter Executives in Line for $100 Million Payout After Elon Musk Fired Them," *The Wall Street Journal*, October 28, 2022, wsj.com/articles/twitter-executives-in-line-for-100-million-payout-after-elon-musk-firing-11666971500.

288 **engaging in collective action:** Josh Eidelson, "Elon Musk's X Illegally Fired Worker Over Protest Tweet, US Labor Board Alleges," *Bloomberg*, October 13, 2023, bloomberg.com/news/articles/2023-10-13/elon-musk-s-x-twitter-illegally-fired-worker-over-protest-tweet-nlrb-alleges.

289 **"One employee's cartoon chronicles Twitter's accelerated descent":** Manu Cornet, Twitter post, October 11, 2023, 11:36 a.m., twitter.com/lmanul/status/1712129939380801800.

289 **"without whose puerile mischief this book would have been very boring":** Manu Cornet, Twitter post, August 21, 2023, 6:58 p.m., twitter.com/lmanul/status/1693804917729722535.

289 **"Collective action seems important now":** Manu Cornet, "'I Thought I'd Been Hacked. It Turned Out I'd Been Fired': Tales of a Twitter Engineer," *The Economist*, December 2, 2022, static.ma.nu/about/aboutme/2022-12-02_economist.png.

289 **"I got 100 million":** Manu Cornet, *Twittoons*, July 17, 2023, twittoons.com/73.

293 **In November 2022, Musk shared:** Elon Musk, Twitter post, November 21, 2022, 8:28 p.m., twitter.com/elonmusk/status/1594865247323660290.

293 **That same month, X's global traffic:** Sarah Perez, "One Year Post-Acquisition, X Traffic and Monthly Active Users Are in Decline, Report Claims," *TechCrunch*, October 17, 2023, techcrunch.com/2023/10/17/one-year-post-acquisition-x-traffic-and-monthly-active-users-are-in-decline-report-claims/.

293 **Monthly ad revenue:** Sheila Dang, "US Ad Revenue at Elon Musk's X Declined Each Month Since Takeover-Data," Reuters, October 4, 2023, reuters.com/technology/us-ad-revenue-musks-x-declined-each-month-since-takeover-data-2023-10-04/.

294 **"the company already is cash flow positive":** Aisha Counts, "X to Test Three Tiers of Service in Latest Musk Overhaul," *Bloomberg*, October 5, 2023, bloomberg.com/news/articles/2023-10-05/x-tells-bankers-that-advertisers-are-back-but-spending-less.

294 **"1000X harder to manipulate the platform":** Elon Musk, Twitter post, October 17, 2023, 5:24 p.m., twitter.com/elonmusk/status/1714437136575807979.

294 **"Since the acquisition, The @ADL":** Elon Musk, Twitter post, September 4, 2023, 1:45 p.m., twitter.com/elonmusk/status/1698754179148214495.

294 "If they lose the defamation suit": Elon Musk, Twitter post, September 4, 2023, 10:59 a.m., twitter.com/elonmusk/status/1698757795259060562.

295 Now there was more hate speech than ever: Sheera Frenkel and Kate Conger, "Hate Speech's Rise on Twitter Is Unprecedented, Researchers Find," *The New York Times*, December 2, 2022, nytimes.com/2022/12/02/technology/twitter-hate-speech.html.